GEOMETRIC DESIGN PROJECTS FOR HIGHWAYS

An Introduction

SECOND EDITION

J.G. Schoon

American Society of Civil Engineers
1801 Alexander Bell Drive
Reston, Virginia 20191–4400

Abstract: This book provides an introduction to geometric design of highways by means of examples and projects that emphasize basic specifications, approaches to preliminary route selection, alignment, drainage, cost, and environmental concerns. Intended as a supplementary text to standard texts on highway engineering for undergraduate and post-graduate university courses, it presents projects from the initial provision of a topographic map and specifications through to the investment and user cost estimates of a particular highway. The ability to connect the various aspects of highway geometric design in terms of a complete project is stressed to assist students and practitioners to understand the design linkages inherent in the design process related to topography and design policy. While intended primarily for university instruction at undergraduate and graduate levels, the book will also be of benefit to transportation and land-use planners wishing to become familiar with the major features of geometric design as it relates to other forms of infrastructure development.

Library of Congress Cataloging-in-Publication Data

Schoon, J.G. (John George), 1937–
    Geometric design projects for highways: an introduction / by J.G. Schoon.—2nd ed.
        p. cm.
    Includes bibliographical references and index.
    ISBN 0-7844-0425-9
    1. Highway engineering. 2. Roads—Design and construction. I. Title.

TE145 .S36   1999
625.7—dc21   99-054577

Contents

List of Figures and Tables

FIGURES

TABLES

Preface

The main purpose of this book is to assist in consolidating the many elements of highway design and linking them into a route selection and geometric design project. This second edition is based upon metric units of measurement. In addition, it enlarges upon environmental reporting concerns and presents a discussion of economic cost analysis and its application. The latter will assist in comparing different projects conducted in a class setting, and is intended to add further realism to the overall design and evaluation process. Also added are the main features of route selection and design aided by digital terrain and computerized alignment modeling. This latter approach is becoming more prominent as its cost is reduced and experience is gained in its use by highway agencies and consulting firms.

Intended for use by senior undergraduate students in civil engineering and graduate students who require a basic highway design course, the book is structured to complement highway design theory described in existing texts and design guidelines, and to supplement these in a typical highway design course. The book is also intended to assist an introductory short-course on geometric design for practicing engineers. It is assumed that the student has a working knowledge of geometry, trigonometry, soil mechanics, hydraulics, and surveying principles. These are subjects which most undergraduate civil engineering students have studied during or before their senior year.

Understanding the interrelationship between geometric design and topography is a fundamental requirement in highway engineering, for these essential elements establish the horizontal and vertical alignment of the centerline, upon which all other details of the highway and the right-of-way depend. The principles remain the same for highways ranging from a simple, two-lane, local road to a multi-lane freeway. Also, the design of any highway route is a unique undertaking in that detailed features of the terrain and other environmental conditions invariably differ; only by working through a practical example which contains the essential design elements can the student be sure of understanding the problems involved, and of developing realistic solutions.

The examples illustrate the process of conducting a preliminary highway design based upon the geometric design controls, topographic maps, and, where possible, aerial photography. Thus, the examples provide exposure to and some practice in determining how the technical feasibility of alternative routes may be verified in terms of the initial location of the proposed highway's centerline and its major physical features in order to:

1. determine a reasonably accurate cost estimate for a technically feasible highway project, on which to base a preliminary economic analysis; and

2. enable a more detailed survey to be made along the defined centerline to verify topographic features, explore in more detail the soil and subsurface conditions, identify key environmental concerns, and make any necessary adjustments before a more detailed design of the highway is undertaken.

Figure 1 outlines the main features of the implementation process to illustrate where a "preliminary" highway design lies within a state or other official agency's activities. The emphasis in the book is to illustrate how the design guidelines (Item 1) and site characteristics (Item 2) together help to establish the design controls (Item 3) as a basis for the preliminary route selection and design (Items 4 and 5), and how the evaluation of alternatives (Item 6) may result in a need to modify the design controls (Item 8). The greatest emphasis will be on the route selection, geometric design, and economic cost estimation processes of Items 4, 5, and 6. Final evaluation and designs (Items 8 and 9, respectively) are not addressed in detail, and invariably involve refinements to the preliminary design. An example of a highway implementation process is described further in Chapter 1.

This book concentrates on design of rural highways. This is because rural highways typically present more extensive problems of earthworks and fitting the alignment into a natural, often mountainous terrain than do urban highways. Urban highways are typically constrained by property lines and existing rights-of-way which often limit the opportunity to explore the implications of alternative alignments Also, except for limited access highways, lower speed limits imposed in the urban setting often do not pose the problems of curvature and its coordination experienced in rural environments, although they may have unique problems of their own, particularly intersection design and traffic control. Many excellent texts deal with these matters and several are mentioned in the bibliography.

Although the design process may vary somewhat depending on each state's or other organization's procedures (for example, in many cases, detailed, large-scale surveys rather than USGS topographic maps may be used), Figure 1 typifies the sequence of activities.

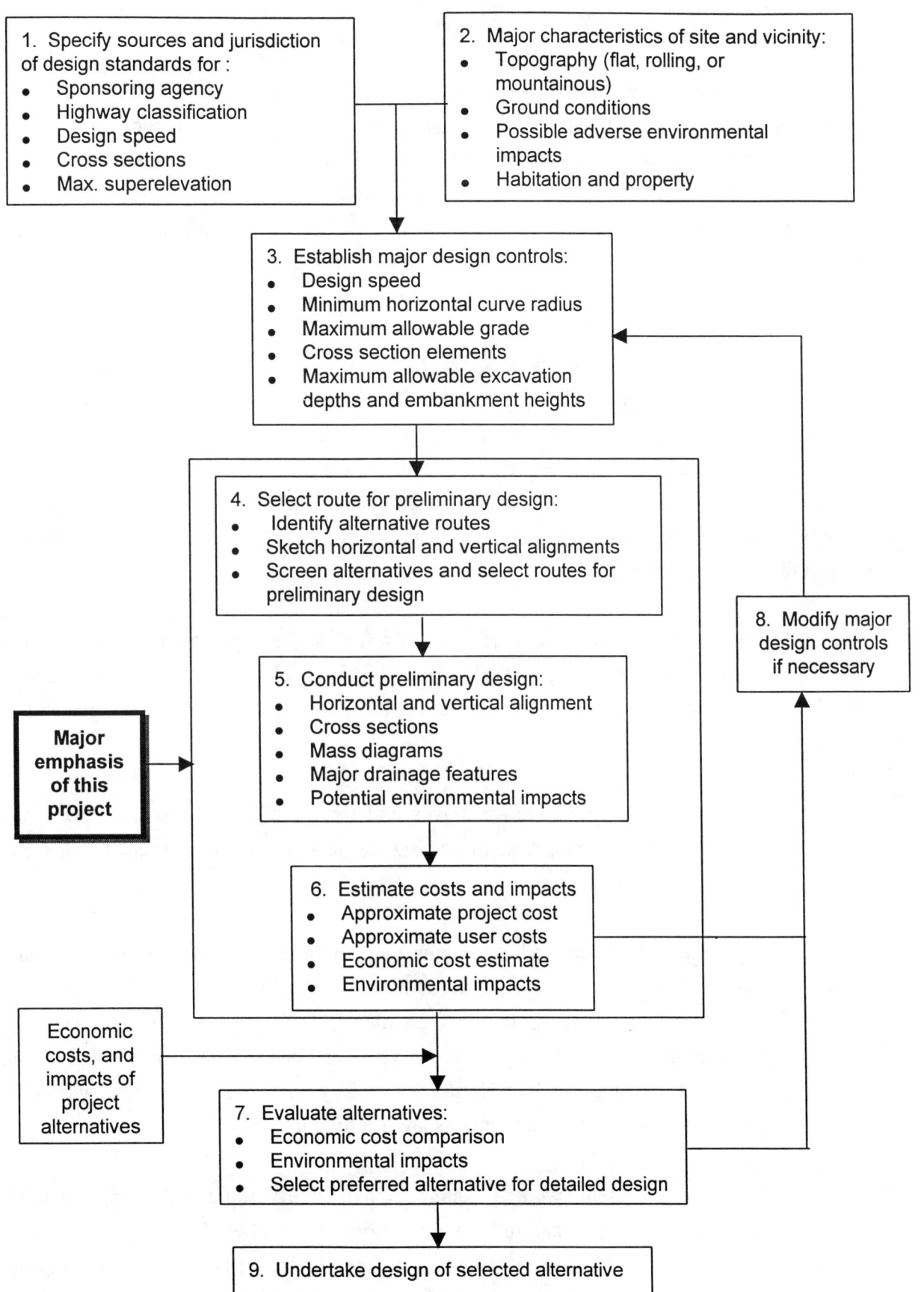

FIGURE 1

HIGHWAY GEOMETRIC DESIGN PROCESS – MAJOR FEATURES

In addition, the examples of preliminary highway design shown in this book may typify highways required as an ancillary component of a larger project, such as an access road to and from a residential or industrial site development. Here, the engineer's concern may focus on the need for a preliminary design to determine the highway's feasibility and cost as part of the total project. Also, the objective of the design may be the realignment of an existing highway, in which case, the principles described remain the same but greater emphasis may be given to the connections with the original route.

A highway improvement may be <u>technically</u> <u>feasible</u> yet not <u>economically</u> <u>justifiable</u> or <u>environmentally acceptable.</u> The only way to determine economic justification and environmental acceptability is to conduct the preliminary design and, based upon expected costs, benefits, and likely environmental impacts, conduct an economic and environmental evaluation. This may result in the preliminary project proceeding to a detailed design stage, being modified and re-evaluated, or being dropped from further consideration.

The book presents the basic objectives of a highway design, the design guidelines, and the methods of establishing a feasible route. This is followed by a worked example and several projects for students to undertake. The main features of each chapter are as follows:

- <u>Chapter 1</u> describes the objectives and the principles of route selection and some of the major factors in inspecting the topography by means of maps and aerial photography.

- <u>Chapter 2</u> presents and comments upon highway design control policy and economic cost estimation generally as described by the American Association of State Highway and Transportation Officials (AASHTO) guidelines, together with certain elements of excavation and fill requirements and drainage design which will affect the highway's technical feasibility and cost. Federal and state environmental laws and requirements are briefly outlined.

- <u>Chapter 3</u> presents worked examples of individual route problems, indicating selection, examination, and refinement of horizontal and vertical alignment related to topographic features. Key route design factors related to environmental impacts are also outlined.

- <u>Chapter 4</u> presents an actual example of the preliminary geometric design and economic cost estimate of a two-lane rural highway as it might be presented and conducted as an individual's project during a typical undergraduate or introductory graduate course in highway engineering. The main steps -- similar to those in the project design -- are illustrated for an example of the computerized digital terrain and alignment modeling process.

- <u>Chapter 5</u> provides several geometric design projects for use by students. End points of potential highways are provided on topographic maps to offer a selection of projects. The instructor may vary these or specify particular routes.

- The <u>Appendix</u> includes map symbols, enlarged maps for students' projects, and design details which may be useful in conducting the projects.

It is emphasized that the projects presented in this book are intended for <u>preliminary</u> analysis only. <u>Detailed engineering analysis of the key features, including safety considerations, will be required before an acceptable design for construction purposes is proposed</u>. Also, although the AASHTO policy on geometric design is used in this book, other policies may be substituted to make the project more representative of the needs of specific locations and conditions. The design policy for low-volume roads is an example of the latter.

The use of microcomputer spreadsheets in some numerical examples has been made for convenience only, and the use of manual methods for the analyses described, using the same basic formats, will not be excessively time-consuming using a pocket calculator. The use of computer programs other than spreadsheets is not addressed in this book because the intent is to familiarize students with the direct relationship between computations and the design product, and this is often not possible with proprietary computer software. Students may prepare their own software, if desired, and the use of commercial programs may be instructive once the principles of the design are understood. The example provided of the digital terrain model and the associated computerized design process will provide some indications of the capabilities of the more automated design aids.

Some ability in technical drawing will be found useful in undertaking the projects. This is because exploration of the spatial relationships between a natural terrain and geometrically definable form are key elements of route selection. Also, the ability to clearly indicate the key points of a design to others is an essential element of the total design process.

I would like to acknowledge the many helpful comments and suggestions made by students and other friends and colleagues in the preparation of this book. Among consulting engineers, James Avitable, Charles Brackett, Walter Herrick, Raymond Oro, Robert Snowber, James Vears, and John Finlayson were particularly helpful. Regarding computerized digital terrain and alignment modeling, David Gianngrande and Suhail Owies provided useful information for the example shown, and Robert Albee, Joseph D. Magni and Colin Munz provided additional information. From an academic perspective for both editions, Dr. B. Kent Lall of Portland State University, Dr. John Mason of Penn State University, Dr. Edward S. Neumann of the University of Nevada at Las Vegas, and Dr. Bob L. Smith of Kansas State University were all very helpful in taking the time to review the book in detail and in making comments on this second edition. All have used the book for elements of their highway design courses. Dr. Ali Touran has used the book to guide students' projects in the design elements of highway engineering coursework and contributed significantly to the material on construction costs. Both he and Dr. Peter Furth made helpful comments regarding the format of the book. Dr. David Navick assisted considerably in collaborating in the economic analysis methodology and in outlining potential routes and design details. Thanks are also due to the civil and environmental engineering departments at Northeastern University, Boston, Massachusetts, U.S.A., and at Southampton University, England, where the assistance of personnel and colleagues alike has been outstanding. In the publishing process, the encouragement and prompt help of ASCE's publishing division has brought the book to fruition, and the assistance and suggestions of Zoe Foundotos, Shiela Menaker, Mary Grace Stefanchik, Lisa Ehmer, Joy Chau, and the editing staff is much appreciated.

Finally, I would like to state that despite all of the assistance rendered, all errors and omissions that may have occurred are mine, and I would be most grateful if readers would tell me about them, together with suggestions for possible improvements.

John G. Schoon
Transportation Research Group
Department of Civil and Environmental Engineering
University of Southampton
Hampshire SO17 1BJ, England   August 1999

Chapter 1

Factors Affecting Selection of the Highway Route

In this chapter, we briefly review some of the main factors affecting selection of the route that can be discerned from available maps, photographs, and other sources. The basic philosophy underlying the relationship between costs and design levels is described to provide some perspective on the appropriateness of specific routes.

A major determinant of the route in a system-wide context is the estimated traffic volume. However, this and highway systems planning in general are beyond the scope of this book. Therefore, the discussions assume that the traffic volumes, vehicle classification, and end points of the proposed highway have been decided as part of a system-wide plan.

EXAMINATION OF NATURAL AND MAN-MADE FEATURES

Selection of a possible route for a proposed highway is -- apart from traffic considerations -- determined largely by relating topographic features, human habitation, and environmental features of the area under consideration to geometric design controls. Therefore, before starting the route selection process, a review of the area's major topographic and other features likely to affect the route selection is needed. Several sources of information are available to assist the review. They include:

1. <u>Topographic Maps</u>. Usually the maps of the United States Geological Survey (USGS) at a scale of 1:25,000 and a contour interval of 3 m provide the minimum required detail for preliminary route selection. Larger scale maps may also be used if available.

2. <u>Aerial Photographs</u>.   The two main types of aerial photographs employed in route design are stereographic and oblique.   Stereographic photographs, usually at the same scale as the topographic maps, may assist in determining important geological, ecological, and cultural information. Also, with the assistance of appropriate measuring devices, they form the basis for automated route selection and design, including the use of computer-aided methods.   Oblique aerial photographs may be used to supplement the stereophotographs.

3. <u>Geological and Soil Maps</u>.   These are available through the USGS, the U.S. Soil Conservation Service, or through state agencies, and may provide useful information, particularly concerning pavement design, although more detailed information is usually necessary for preliminary design.

4. <u>Ground Surveys</u>.   Reconnaissance or more detailed surveys should be made of the area, especially if the terrain is rugged or if additional details are required. As early as possible in the process, the design engineer should "walk the route," but practicalities may preclude this procedure early in the project.

In this book, we will rely primarily upon the information that can be obtained from USGS topographic maps; in practice, the designer may use a combination of several sources of information. The items used and the major steps in the review may typically include the following:

Topographic Maps -- Examine the terrain in general between the start and end points of the proposed route and make note of the following information, usually available from a topographic and geological maps. Typical kinds of information shown on most topographic maps are illustrated in the Appendices. An inspection of the maps should include the following steps:

1. Identify unsuitable ground conditions such as wetlands, rock outcrops, areas subject to flash floods or avalanche, and other features of an obviously difficult terrain for highway construction.

2. Examine the contour lines to obtain an initial estimate of the gradients that exist on undulating or mountainous parts of the potential route. The steepness of the terrain may be approximately determined by observing the number of contour lines and their vertical interval along a horizontal distance located at right angles to them. Slopes steeper than, say, approximately 10%, may be delineated on the map.

3. Define streams, rivers, ravines, or other topographic features that indicate the possible need for bridges or other extensive ancillary works to the highway itself.

4. List typical types of subsurface and soil conditions that may be expected, as indicated by the topographic features found on the topographic and geological map.

5. Summarize the findings of the examination of the above items on maps or overlays in order to guide the next steps in the route selection. Items 1 through 4 are described graphically in Figures 1-1 through 1-3 where, as an example, the route of a highway connecting points A and B is being considered.

Other features that may be identified at this time, and that may be directly related to the effectiveness of the route, include consideration of sunlight availability to reduce the effects of snow and ice accumulations, avoidance of possible avalanche areas, and the effects of the route on habitation and other cultural activities such as schools and community centers. The physical characteristics defined by these considerations can significantly affect the alignment of the route and its ultimate benefits to local and wider communities.

Aerial Photography -- The next step is to examine available aerial and other photography, when available, to confirm or modify the information on the map. The following procedure is usually appropriate:

Examine the stereoscopic aerial photographs to determine whether topographic and cultural features are different from those shown on the map. Document any changes on an overlay so this may be recorded on the map. The type of features found might include human activity, swamp or marsh areas that no longer exist, or areas that may be sensitive

3

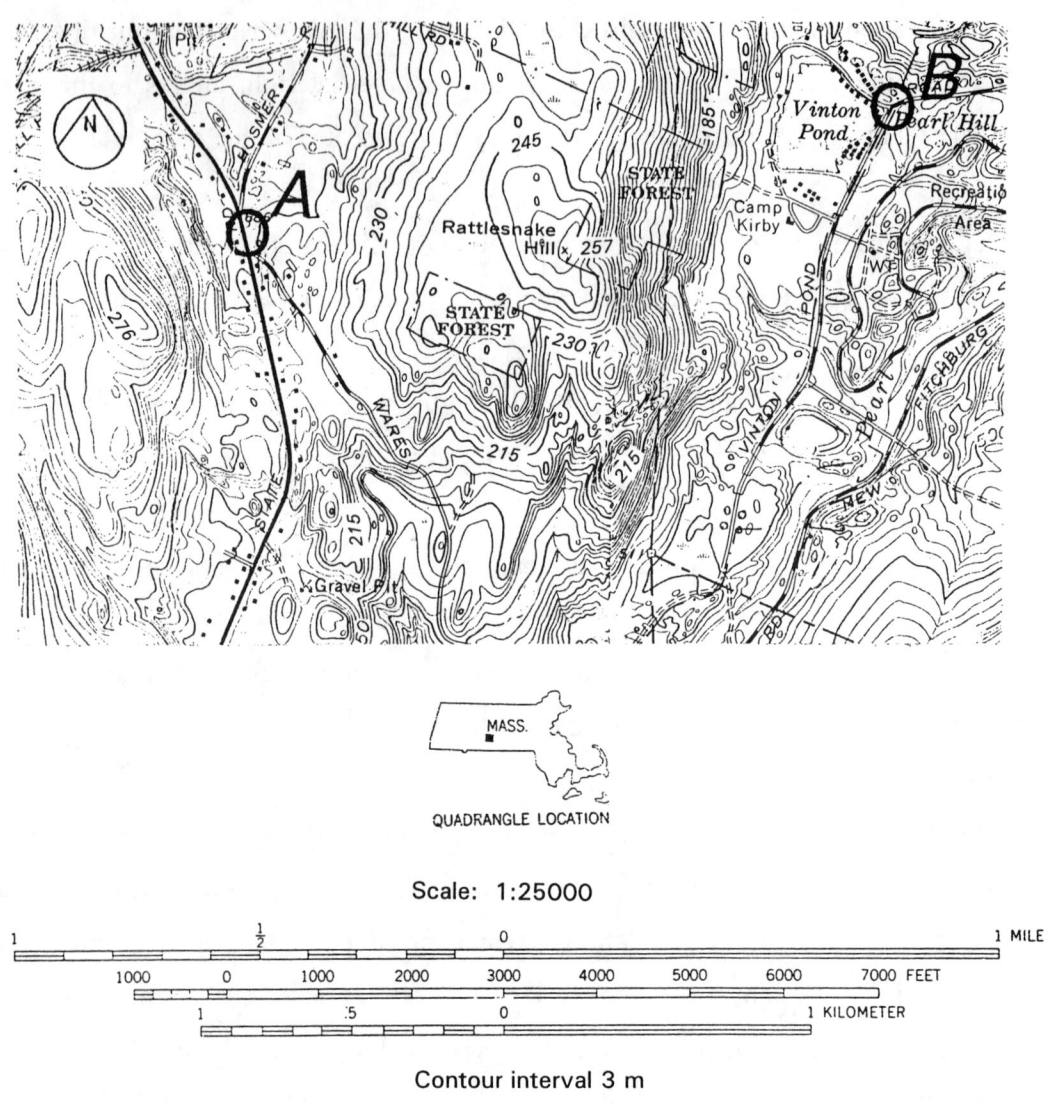

QUADRANGLE LOCATION

Scale: 1:25000

Contour interval 3 m

End points of proposed highway,   A   B

FIGURE 1-1

SELECTED AREA OF INTEREST

Source: Overlay of Figure 1-1

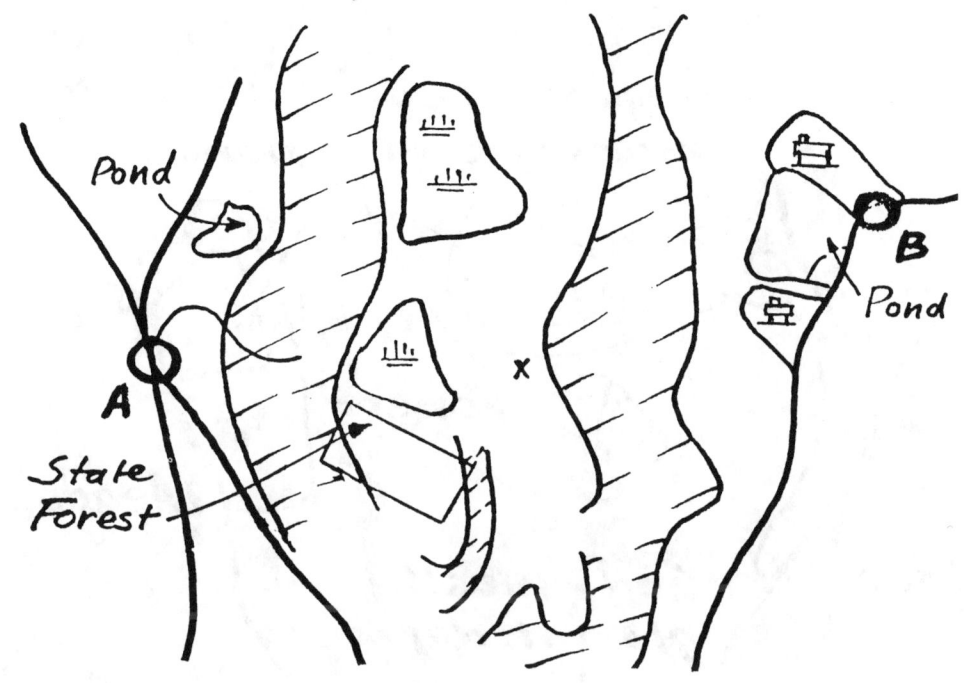

Pond

A

State Forest

B

Pond

LEGEND:
Roads
Marsh
Steep areas
High elevations
Habitation

FIGURE 1-2

EXAMPLES OF TOPOGRAPHIC AND CULTURAL FEATURES

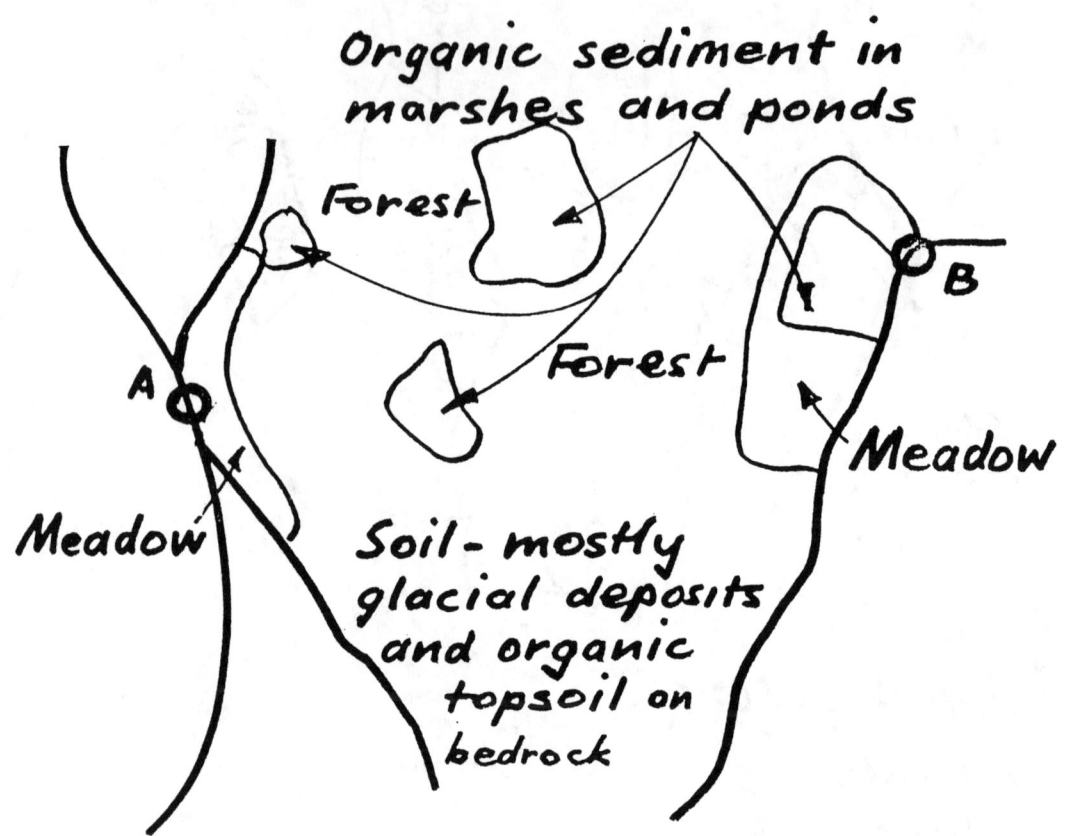

Organic sediment in marshes and ponds

Forest

Forest

A

B

Meadow

Meadow

Soil - mostly glacial deposits and organic topsoil on bedrock

Source: Overlay of Figure 1-1

FIGURE 1-3

EXAMPLES OF SOIL AND VEGETATION FEATURES

due to presence of wildlife or other ecological factors. In addition, conditions potentially hazardous to a highway such as avalanches, mudslides, or flooding may be evident. Aerial photography can provide excellent indications of anticipated ground conditions. Examples of the stereoscopic photographs are presented in Figure 1-4.

Examine the oblique photographs to obtain a sense of the developmental and esthetic features of the area and a general idea of the grades and other topographic characteristics. These, of course, should be cross-checked with those on the map to ensure correspondence. Figure 1-5 provides several oblique photographs of the project area.

Note in particular the presence of trees that may make identification of the ground surface features difficult on the aerial photographs.

In addition to the above considerations, make note of local features which may be environmentally sensitive to the presence of a proposed highway. Guidelines for identifying these features and mitigating the effects of potential highways on the environment are described in several documents listed in the bibliography.

IDENTIFICATION OF TECHNICALLY FEASIBLE ROUTES

The guiding principle in designing a possible route is to improve the transportation between specified points. Within the economic and social framework that typically applies, the term "improve" may be broadly interpreted as "to make less expensive and safer for the public in general as well as for the highway's users, while at the same time maintaining or contributing to the improvement of environmental quality." Furthermore, the route should be "technically feasible" in that no excessive construction or maintenance problems are envisaged, and such that the design controls and policy on geometric design of the highway agency having jurisdiction are adhered to. In this book, the policies of the American Association of State Highway and Transportation Officials (AASHTO) are generally used.

A highway improvement may be an upgraded existing highway or a completely new route and should always be considered as a component of the overall transportation system.

7

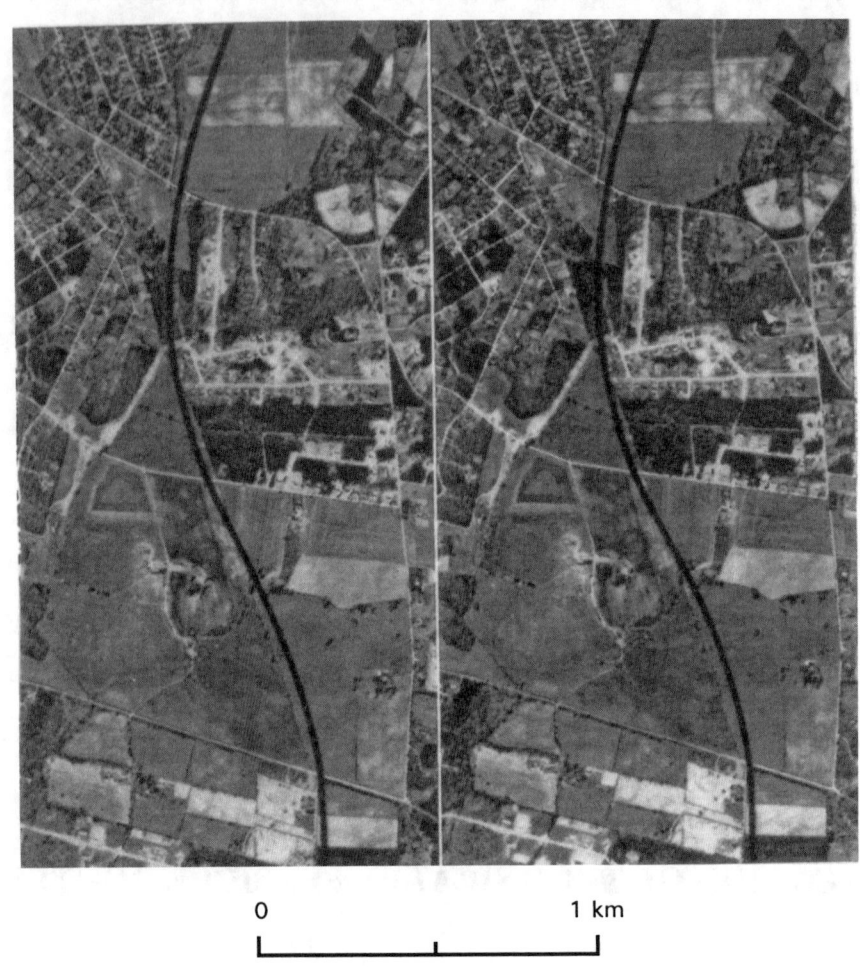

0   1 km

Source: Belcher, Donald J. "Photo Interpretation in Engineering." Manual of Photographic Interpretation. Ed. Robert N. Colwell. Washington, DC: American Society of Photogrammetry, 1960. page 425.

FIGURE 1-4

EXAMPLES OF STEREOGRAPHIC AERIAL PHOTOGRAPHS

Source: Federal Highway Administration, "I-70 In a Mountain Environment"
Report No. FHWA-TS-78-208, Washington. D.C., 1978.

FIGURE 1-5

EXAMPLES OF OBLIQUE
AERIAL PHOTOGRAPHS

OBJECTIVES IN IDENTIFYING ACCEPTABLE ROUTES

In defining a broadly acceptable route, therefore, the approach typically involves compromising between the user costs and construction costs while seeking the route and physical conditions that result in the least adverse environmental impact.

How is a balance struck between user costs and construction costs? A rather extreme example may be used to illustrate this problem: suppose the objective is to define a route between two points on existing highways separated by mountainous terrain. The least cost route for vehicle <u>users</u> on a "per vehicle kilometer" basis would clearly be a horizontal and vertical alignment permitting a high design speed (long sight distances, large radius curves, etc.) route with bridges and tunnels and extensive cuts and fills to overcome the rugged terrain. At the other extreme, a winding road following the contours of the terrain, with little or no cut and fill sections, few bridges, and no tunnels, would result in higher user costs due to sharp curves, resultant reductions in speed, and greater likelihood of accidents. However, such a road would undoubtedly cost less to build, even if it were somewhat longer than the first, because of the reduced amount of expensive excavation and filling and construction of bridges and tunnels.

In a more formalized way, as developed by the World Bank, the Highway Design and Maintenance Model states that (in selecting a particular highway) " ...the basic task is to predict total life-cycle costs – construction, maintenance, and road user costs – as a function of the road design, maintenance standards and other policy options which may be considered", and adds that a broader definition of societal costs would include such examples as air pollution as it affects non-road-users.

For any given volume of traffic, the relationship between the user cost, construction cost, and total cost can be shown conceptually, as in Figure 1-6. The lower the design standards, the lower the construction costs (because of the reduced need for cut, fill, bridges, and tunnels). Conversely, the travel cost to users increases due to the reduced speed and increased travel time, and the increased likelihood of accidents due to the lower geometric design standards. Examples of higher and lower geometric design standards

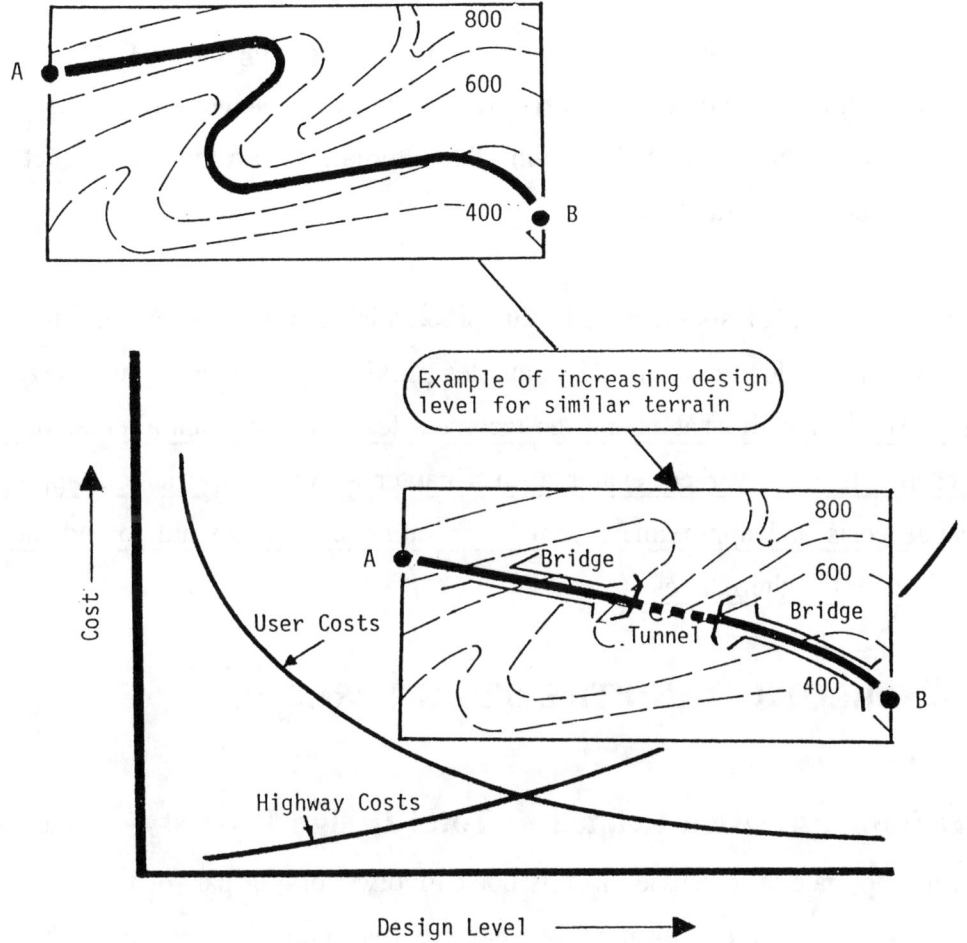

FIGURE 1-6

COST AND DESIGN LEVEL RELATIONSHIPS

applied to highways are shown in Figure 1-7. It must be emphasized that both of these highways satisfy specified criteria in terms of their function and role within the overall system, and the terms "higher" and "lower" should not be construed as meaning "better" and "worse" designs.

The preferred route can only be determined by comparing the total costs for users and the construction and maintenance costs incurred by the implementing agency, for each technically feasible alternative, and selecting that alternative with the least monetary cost and acceptable non-quantifiable impacts.

The economic analysis involved in this process is described in several publications. See the bibliography for these. In the examples provided in this book, however, we are concerned primarily with establishing the technical feasibility of each alternative, and its capital cost. In the worked example in Chapter 4 we also include approximate maintenance costs and approximate vehicle operating cost as an aid to indicating the relative merits of the alternatives.

ROUTE SELECTION AND THE DESIGN PROCESS

Preliminary Design Related to Total Design Process -- As indicated in the Preface, the material discussed in this book involves only a part of the full highway design process, and each organization will have its own guidelines and procedures to be adhered to.

An example process is that for a state highway design in Massachusetts, as shown in Table 1-1. A total of 55 steps are required; Nos. 16 through 22 typify the activities of the highway designer as described in this book. The next step, No. 23, enables the 25% project review to be conducted. In this step the preliminary design and cost estimates are reviewed by federal and state officials. This ensures that any problems in the project can be adequately identified and addressed and that the project is sufficiently advanced to proceed with public hearings and the subsequent permitting, detailed design, and approval process.

Lower
design
standard

Higher
design
standard

FIGURE 1-7

EXAMPLES OF HIGHWAYS BUILT TO DIFFERENT DESIGN STANDARDS

Source: U.S. Department of Transportation, Federal Highway Administration

13

TABLE 1-1
EXAMPLE OF HIGHWAY DESIGN PROCESS STEPS

01	DOCUMENT ROAD IMPROVEMENT NEED
02	PROJECT REVIEW COMMITTEE ACTION
03	PROJECT INCLUDED IN THE TRANSPORTATION IMPROVEMENT PROGRAM (TIP)
04	MDPW INITIATES PROJECT
05	COMPILE EXISTING PROJECT DATA
06	WALK THE PROJECT
07	REFINE PROJECT LIMITS
08	DETERMINE MEPA AN NEPA PROJECT CATEGORY
09	DETERMINE OTHER APPLICABLE FEDERAL, STATE, AND LOCAL ENVIRONMENTAL LAWS AND REQUIREMENTS
10	DETERMINE PUBLIC PARTICIPATION REQUIREMENTS
11	BEGIN INTERAGENCY COORDINATION
12	PROCESS ENVIRONMENTAL DOCUMENT
13	HOLD PUBLIC HEARINGS
14	ORDER NECESSARY SURVEY DATA
15	REDUCE AND PLOT SURVEY DATA

16	REQUEST NECESSARY TRAFFIC DATA
17	DEVELOP HORIZONTAL AND VERTICAL GEOMETRICS
18	DEVELOP TYPICAL CROSS SECTIONS
19	DEVELOP PRELIMINARY RIGHT-OF-WAY PLANS
20	DEVELOP DRAFT TRAFFIC SIGNAL PLAN (IF REQUIRED)
21	DEVELOP BRIDGE TYPE STUDIES AND SKETCH PLANS FOR BRIDGES, CULVERTS, AND WALLS (IF REQUIRED)
22	DEVELOP PRELIMINARY COST ESTIMATE

23	CONDUCT 25% PROJECT REVIEW
24	OBTAIN 25% PROJECT APPROVAL
25	START PERMIT PROCESSES
26	REVIEW PROJECT CHANGES FOR MEPA PURPOSES
27	INTERAGENCY COORDINATION
28	REVIEW PROJECT CHANGES FOR NEPA PURPOSES
29	COMPLETE PERMIT PROCESSES
30	COMPUTE HORIZONTAL AND VERTICAL GEOMETRY
31	PREPARE SUBSURFACE EXPLORATORY PLAN (IF NECESSARY)
32	DEVELOP CROSS SECTION TEMPLATES
33	DEVELOP CONSTRUCTION TRACINGS
34	TRANSMIT PLANS TO RAILROAD-UTILITIES ENGINEER
35	DEVELOP TRAFFIC-RELATED PS&E DATA
36	DEVELOP PAVEMENT DESIGN
37	DEVELOP FINAL DRAINAGE DESIGN
38	DEVELOP PRELIMINARY RIGHT-OF-WAY AND/OR LAYOUT PLANS
39	COORDINATE UTILITY RELOCATIONS
40	UPDATE CONSTRUCTION TRACINGS
41	UPDATE COST ESTIMATE
42	CONDUCT 75% PROJECT REVIEW
43	75% PROJECT APPROVAL
44	REVIEW PROJECT CHANGES FOR MEPA PURPOSES
45	INTERAGENCY COORDINATION
46	REVIEW PROJECT CHANGES FOR NEPA CHANGES
47	DEVELOP TRAFFIC CONTROL PLAN (TCP) THROUGH CONSTRUCTION ZONES
48	DEVELOP TRAFFIC CONTROL AGREEMENT WITH MUNICIPALITY (IF REQUIRED)
49	FINALIZE LAYOUT PLANS AND ORDER RIGHT-OF-WAY
50	FINALIZE RIGHT-OF WAY PLANS
51	FINALIZE CONSTRUCTION PLANS
52	FINALIZE COST ESTIMATES
53	DEVELOP SPECIAL PROVISIONS
54	TRANSMIT CONSTRUCTION PLANS TO RAILROAD UTILITIES ENGINEER
55	PS&E SUBMITTAL

Area of emphasis in this book

16. REQUEST NECESSARY TRAFFIC DATA

The designer must request traffic operational characteristics. Both existing and projected traffic data is obtained. This data includes ADT, peak-hour volumes, directional distribution, K factor, design-hour volumes (DHV), and percentage of trucks.

17. DEVELOP HORIZONTAL AND VERTICAL GEOMETRICS

The designer must develop the basic horizontal and vertical curvature and grade data.

18. DEVELOP TYPICAL CROSS SECTIONS

Based on design requirements, typical cross sections are developed. Typical cross sections show design elements that will predominate throughout the project.

19. DEVELOP PRELIMINARY RIGHT-OF-WAY PLANS

Preliminary right-of-way plans are developed based on horizontal and vertical geometrics, typical cross sections, and visual and environmental/planning considerations.

20. DEVELOP DRAFT TRAFFIC SIGNAL PLAN (If required)

Based upon guidelines provided in the *Manual on Uniform Traffic Control Devices*, a Draft Traffic Signal Plan is developed.

21. DEVELOP BRIDGE TYPE STUDIES AND SKETCH PLANS FOR BRIDGES, CULVERTS, AND WALLS (If required)

The Type Studies are a preliminary presentation of the overall concept of the proposed structure which shows all pertinent details for the preparation of sketch plans and contract plans.

22. DEVELOP PRELIMINARY COST ESTIMATE

Prepare an estimate based on the latest project information. Refinements are to be expected as the design develops, but this estimate should reflect project costs as accurately as they can be defined at the 25% design stage.

Source: Based upon Massachusetts Department of Public Works, Highway Design Manual, 1990.

14

Environmental Reporting Requirements -- The level of detail of the preliminary route selection and design is also consistent with Class III actions in accordance with the Federal Highway Administrations' level of documentation as described in the appropriate code of Federal Regulations, Title 23 Part 771. A Class III action is an action in which the significance of the impact is not clearly established. These actions require the preparation of an Environmental Assessment describing the environmental impacts of the proposed works and its alternatives. This assessment assists in deciding the nature of further environmental analysis and needed documentation. If the Federal Highway Administration determines that a proposed project will not have a significant impact on the environment, a statement to that effect may be prepared. The level of detail of the project described in this book would in many respects be appropriate for guiding the preparation of such an environmental assessment document. Further details of federal and state requirements are described in Chapter 2, examples are presented in Chapter 3, and key issues of importance are briefly noted in the design project of Chapter 4.

Bibliography of Selected Publications

GENERAL TEXTS ON HIGHWAY ENGINEERING

Garber, N.J., and Hoel, L.A. (1987). Traffic and Highway Engineering. St. Paul, MN: West Publishing Co.

Mannering, Fred L., and Kilarski, Walter P. (1990). Principles of Highway Engineering and Traffic Analysis. New York: John Wiley and Sons.

O'Flaherty, C.A. (198_). Highways. Vol. 2. London: Arnold.

Oglesby, C.H. (1975). Highway Engineering. Third ed. New York: John Wiley and Sons.

Salter, R.J. (1982). Highway Design and Construction. Second ed. London: MacMillan Co.

Thagesen, (1996) Bent. Ed. Highway and Traffic Engineering in Developing Countries. London: F. and N. Spon.

Woods, K.B., ed. (1960). Highway Engineering Handbook. New York: McGraw-Hill.

Wright, P.A., and Paquette, R.J. (1996). Highway Engineering. Fifth ed. New York: John Wiley and Sons.

GEOMETRIC DESIGN

American Association of State Highway and Transportation Officials. (1994). "A Policy on Geometric Design of Highways and Streets." Washington, D.C.

American Association of State Highway and Transportation Officials. (1965). "A Policy on Geometric Design of Rural Highways." Washington, D.C.

American Association of State Highway Officials. (1973). "A Policy on Design of Urban Highways and Arterial Streets." Washington, D.C.

Cron, F.W., et al. (1977). Practical Highway Esthetics. New York: American Society of Civil Engineers.

Department of Transport, UK. (1997) "Design Manual for Roads and Bridges. Volume 6, Road Geometry." London: Her Majesty's Stationary Office.

ROUTE LAYOUT AND SURVEYING

Cron, F.W., et al. (1977). Practical Highway Esthetics. New York: American Society of Civil Engineers.

Hickerson, T.F. (1959). Route Surveys and Design. 4th ed. New York: McGraw-Hill.

McHarg, I. (1969). Design with Nature. New York: The Natural History Press.

Meyer, C.F. (1962). Route Surveying. Third ed. Scranton, Pa: International Textbook Co.

ECONOMIC ANALYSIS

American Association of State Highway and Transportation Officials. (1977). "A Manual on User Benefit Analysis of Highway and Bus Transit Improvements." Washington, D.C. (This book is sometimes referred to as the "Red Book").

Campbell, B. and Humphrey, Thomas F. (1988) "Methods of Cost-Effectiveness Analysis for Highway Projects." NCHRP Report 142. Washington, DC: Transportation Research Board.

Dickey, J.W., and Miller, L.H. (1984). Road Project Appraisal for Developing Countries. New York: John Wiley and Sons.

Grant, Eugene l.; Ireson, W. Grant and Leavenworth, Richard S. (1976). Principles of Engineering Economy. New York: The Ronald Press Company.

Thomas, E.N., and Schofer, J.L. (1970). "Strategies for the Evaluation of Alternative Transportation Plans." NCHRP Report 96. Washington DC: National Research Council.

Watanatada, T., et al. (1989). The Highway Design and Maintenance Standards Model. Baltimore and London: Johns Hopkins Universtiy Peess.

Winfrey, R. (1968). Economic Analysis for Highways. Scranton, Pa: International Textbook Co.

AERIAL PHOTOGRAPHY

American Society of Photogrammetry. (1960). Manual of Photographic Interpretation. Washington, D.C.

Paine, D.P. (1981). Aerial Photography. New York: John Wiley and Sons.
Ray, R.G. (1960). Aerial Photographs in Geologic Interpretation and Mapping. Washington, DC: U.S. Government Printing Office.

ENVIRONMENT

McHarg, I. (1969). Design with Nature. New York: The Natural History Press.

"Regulations for Implementing the Procedural Provisions of the National Environmental Policy Act, Council on Environmental Quality." Nov. 29, 1978. Executive Office of the President, Washington, D.C.

American Association of State Highway and Transportation Officials. (1993). "Guide on Evaluation and Abatement of Traffic Noise." Washington, D.C.

Cowardin, L. (1992) "Classification of Wetlands and Deepwater Habitats in the United States. Fish and Wildlife Service." Report FWS/OBS-79/31. Department of the Interior: Washington, D.C.

Skidmore, Owings and Merrill. (1975). Environmental Notebook Series. Prepared for the U.S. Department of Transportation, U.S. Government Printing Office, Washington, D.C.

Wholley, T., ed. (1994). Fundamentals of Air Quality. New York: American Society of Civil Engineers.

DRAINAGE

American Association of State Highway and Transportation Officials. (1992). "Guidelines for Hydraulic Considerations in Highway Planning and Location, Volume 1." Washington, D.C.

Federal Highway Administration. (1965). "Design of Roadside Drainage Channels." Washington DC: Hydraulic Design Series No. 4.

Federal Highway Administration. (1973). "Design Charts for Open Channel Flow ". Washington DC: Hydraulic Design Series No. 3.

Federal Highway Administration/National Highway Institute. (1990). "Highways in the River Environment, Hydraulic and Environmental Considerations, Training and Design Manual." Washington D.C.

U.S. National Weather Service. (1961). "Rainfall Frequency Atlas for the United States." Technical Paper No. 4, Washington, D.C.

Young, Kenneth, and Krolak, Joseph S. (1994). HYDRAIN, Integrated Drainage Design Computer System, Version 5. Developed by GKY and Associates, Inc., Springfield, VA, for the Office of Engineering and Highway Operations, Research and Development, Federal Highway Administration, Washington, D.C.

Chapter 2
Design Controls and Guidelines

A highway design that provides the desired service requires the specification of controlling variables responsive to the intended use of the highway and to its expected economic and environmental features. Therefore, this chapter describes selected major factors associated with design controls and economic cost estimates, mostly as described by the American Association of State Highway and Transportation Officials (AASHTO), and in accordance with relevant environmental reporting requirements. The information is presented in approximately the same order as it might be required in conducting the design project described in Chapter 4. The coverage of the material is therefore selective and is intended to be consistent with this approach.

Again, it is emphasized that no attempt is made here to present the theory underlying the controls or guidelines. The reader should refer to Ref. 1 for design procedures and greater detail and to the texts that deal with these subjects (also see bibliography in Chapter 1).

DESIGN CONTROLS

Once the functional classification (e.g., collector, arterial, etc.) of a proposed highway has been defined, major controls affecting the design must be specified in order to guide the design process. The functional classifications of highways in rural areas include principal arterials, minor arterials, major and minor collectors, and local roads. At a minimum, the following design controls usually must be specified·

Traffic volumes
Design vehicle
Design speed
Maximum grade
Lane and shoulder widths

Each of these items is discussed briefly below. Because Chapter 4 illustrates a design of a <u>rural collector highway</u>, emphasis is given to this highway classification in the examples.

Traffic Volumes --The traffic data should include the current and future average daily traffic (ADT), the future design hourly volume (DHV), directional distribution, and truck percentage, as indicated in Table 2-1.

On most highways, the DHV is used for design. The ADT may be used for design of minor, low-volume roads. On highways with unusual or highly seasonal fluctuations in traffic flow, the DHV should be based upon detailed analysis of the anticipated demand. For important intersections, data should be obtained to show traffic movements during morning and evening peak hours and at other times of heavy traffic.

The capacity of each highway and the levels of service associated with the demand must be determined from a capacity analysis. For design purposes, the guidelines shown in Table 2-2 may provide an initial indication of the appropriate levels of service. For example, if other data are unavailable to indicate otherwise, level-of-service D would be appropriate for a rural collector highway in mountainous terrain. Estimation of actual service flow rates, level of service, and related performance measures should be done in accordance with procedures described in the latest edition of the <u>Highway Capacity Manual</u> (Ref. 4).

Design Vehicles—Width and height, overhangs, and minimum turning paths at intersections are key dimensions to be noted and accommodated. Current policy (Ref. 1) states that the vehicle that should be used in design for normal operation is the largest one that represents a significant percentage of the traffic for the design year. For design of most highways accommodating truck traffic, one of the design semitrailer combinations is typically used. When geometric configuration is constricted, such as in urban areas and at certain intersections, a design check should be made for the largest vehicle expected to ensure that it can negotiate the designated turns, particularly if pavements are curbed. In special cases, a design may have to be made to accommodate vehicles larger than the WB-15. Minimum turning paths are shown in Table 2-3. Templates are typically used to determine the location of wheel paths.

Table 2-1
Traffic elements and their relation - rural highways
(Source: Ref. 2, Table II-6)

Traffic element	Explanation and nationwide percentage or factor
Average daily traffic: ADT	Average 24-hour volume for a given year; total for both directions of travel, unless otherwise specified. Directional or one-way ADT is an average 24-hour volume in one direction of travel only.
Current traffic	ADT composed of existing trips, including attracted traffic, that would use the improvement if opened to traffic today (current year specified).
Future traffic	ADT that would use a highway in the future (future year specified). Future traffic may be obtained by adding generated traffic, normal traffic growth, and development traffic to current traffic, or by multiplying current traffic by the traffic projection factor.
Traffic projection factor	Future traffic divided by current traffic. General range, 1.5 to 2.5 for 20-year period. (Freeways may be up to 20 percent greater or 1.8 to 3.0.)
Design hour volume: DHV	Future hourly volume for use in design (two-way unless otherwise specified), usually the 30th highest hourly volume of the design year (30 HV) or equivalent, the approximate value of which can be obtained by the application of appropriate percentages to future traffic (ADT). The design hour volume, when expressed in terms of all types of vehicles, should be accompanied by factor T, the percentage of trucks during peak hours. Or, the design hour volume may be broken down to the number of passenger vehicles and the number of trucks.
Relation between DHV and ADT: K	DHV expressed as a percentage of ADT, both two-way; normal range, 12 to 18. Or, DHV expressed as a percentage of ADT, both one-way; normal range, 16 to 24.
Directional distribution: D	One-way volume in predominant direction of travel expressed as a percentage of two-way DHV. General range during design hour 50 to 80. Average, 67.
Composition of traffic: T	Trucks (exclusive of light delivery trucks) expressed as a percentage of DHV. Average 7 to 9. Where week-end peaks govern, average may be 5 to 8.

21

	Type of Area and Appropriate Level of Service			
Highway Type	Rural Level	Rural Rolling	Rural Mountainous	Urban and Suburban
Freeway	B	B	C	C
Arterial	B	B	C	C
Collector	C	C	D	D
Local	D	D	D	D

NOTE: General operating conditions for levels of service (Source: Ref. 11):
    A - free flow, with low volumes and high speeds.
    B - reasonably free flow, but speeds beginning to be restricted by traffic conditions.
    C - in stable flow zone, but most drivers restricted in freedom to select their own speed.
    D - approaching unstable flow, drivers have little freedom to maneuver.
    E - unstable flow, may be short stoppages.

Table 2-2. Guide for selection of design levels of service (Source: Ref. 1, Table II-6)

Design Vehicle Type	Passenger Car	Single Unit Truck	Single Unit Bus	Articulated Bus	Semi-trailer Intermediate	Semi-trailer Combination Large	Semi-trailer Full Trailer Combination	Inter-State Semi-Trailer	Inter-State Semi-Trailer	Triple Semi-Trailer	Turnpike Double Semi-Trailer	Motor Home	Passenger Car with Travel Trailer	Passenger Car with Boat and Trailer	Motor Home and Boat Trailer
Symbol	P	SU	BUS	A-BUS	WB-12	WB-15	WB-18	WB-19*	WB-20**	WB-29	WB-35	MH	P/T	P/B	MH/B
Minimum design turning radius (m)	7.3	12.8	12.8	11.6	12.2	13.7	13.7	13.7	13.7	15.2	18.3	12.2	7.3	7.3	15.2
Minimum inside radius (m)	4.2	8.5	7.4	4.3	5.7	5.8	6.8	2.8	0	6.3	5.2	7.9	0.6	2.0	10.7

* Design vehicle with 14.6 m trailer as adopted in 1982 STAA (Surface Transportation Assistance Act).
** Design vehicle with 16.2 m trailer as grandfathered in 1982 STAA (Surface Transportation Assistance Act).

Table 2-3. Minimum turning radii of design vehicles (Source: Ref. 1, Table II-2)

Design Speed Designation -- The design speed is a primary determinant of the geometric design. In order to provide better selection of potential physical design values, a distinction has been made between "rural highways and high-speed urban streets," and "low-speed urban streets" (Ref. 1). The former category is used in the examples shown later in this book. However, the latter may be appropriate for urban street designs where smaller radii and lower coefficients of friction are appropriate. As well as the selected design speed, the design will be dependent upon the traffic, highway capacity, and running speed and will reflect the following variables:

Terrain -- As the terrain varies from level to mountainous, so the cost of the construction for any given speed will increase. Although difficult to define precisely, examples of level, rolling and mountainous are shown graphically in Figure 2-1. Definitions of terrain as related to highway design are given in Ref. 1 as follows:

"Level terrain is that condition where highway sight distances, as governed by both horizontal and vertical restrictions, are generally long or could be made to be so without construction difficulty or major expense.

"Rolling terrain is that condition where the natural slopes consistently rise above and fall below the road or street grade and where occasional steep slopes offer some restriction to normal horizontal and vertical roadway alignment

"Mountainous terrain is that condition where longitudinal and transverse changes in the elevation of the ground with respect to the road or street are abrupt and where benching and side hill excavation are frequently required to obtain acceptable horizontal and vertical alignment.

"Terrain classifications pertain to the general character of a specific route corridor. Routes in valleys or passes of mountainous areas that have all the characteristics of roads or streets traversing level or rolling terrain should be classified as level or rolling. In general, rolling terrain generates steeper grades, causing trucks to reduce speeds below those of passenger cars, and mountainous terrain aggravates the situation, resulting in some trucks operating at crawl speeds."

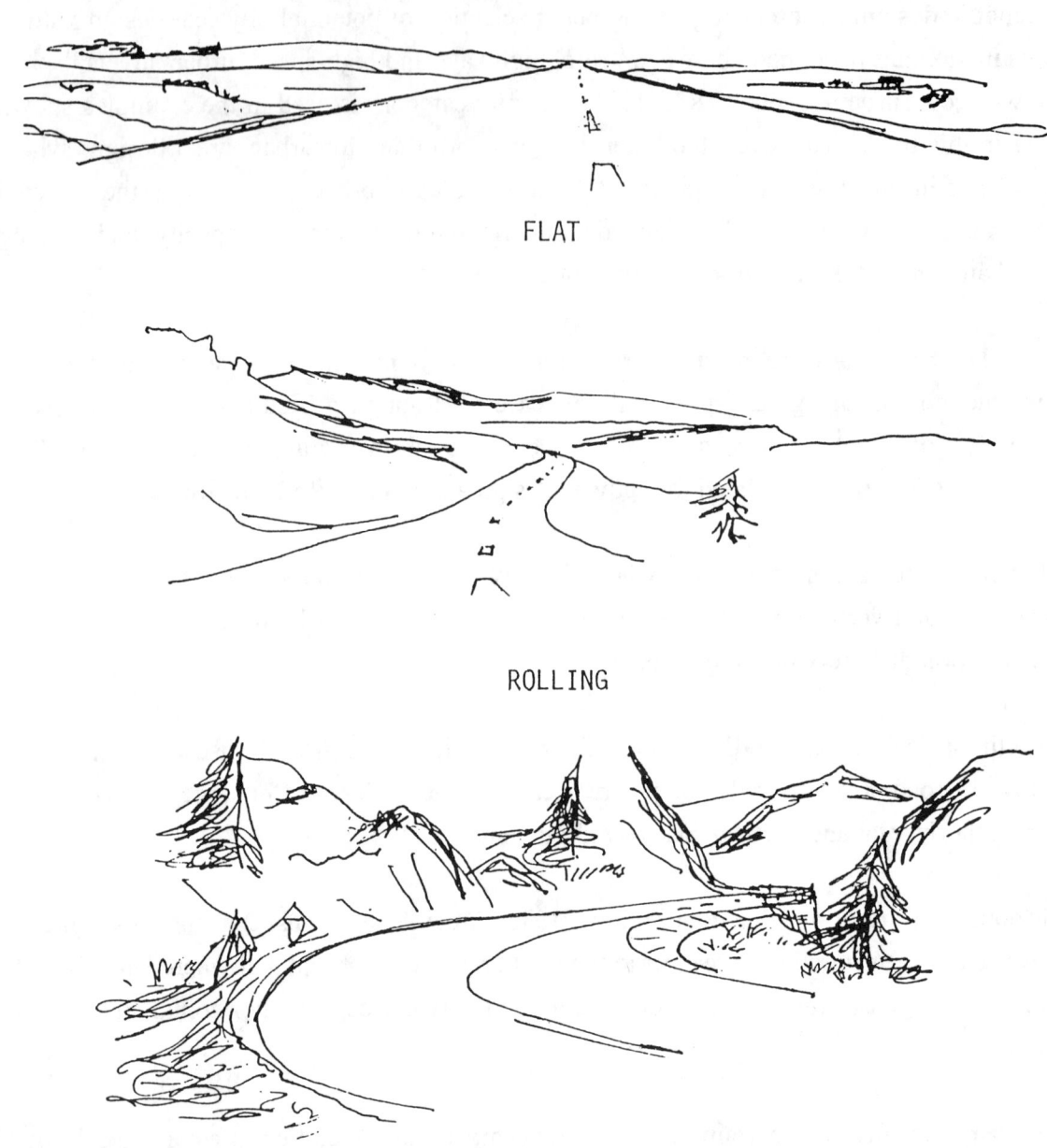

FLAT

ROLLING

MOUNTAINOUS

FIGURE 2-1
EXAMPLES OF TERRAIN CATEGORIES

Grades -- The higher the maximum grade, the lower will be the design speed, reflecting the lower running speeds of vehicles using the highway, and commercial vehicles in particular.

Lane and Shoulder Widths -- Higher values of lane increase design capacity. Designated lane widths range between 3 m and 3.6 m for highways with any significant traffic volumes. Adequate shoulders are usually necessary for safety and capacity reasons. Example: Selection of Design Controls -- The relationships of the above variables and design capacities are assembled in Tables 2-4, 2-5, and 2-6 for rural collector highways The design speed may be selected from these tables. For example, given the following data, determine an appropriate design speed, maximum grade, lane width, and shoulder widths:

Terrain: mountainous
ADT: 3500

Table 2-4 indicates that a minimum design speed of 60 km/h would be required, and Table 2-5 indicates a maximum grade of 10% for that design speed. The appropriate lane width would be 3.6 m with 2.4 m shoulders (Table 2-6). It should be noted that higher speeds may be justified if safety and cost considerations are adequately met.

Design Designation—In order to summarize and present the information on design controls and criteria, it is usual to indicate on the title sheet of the set of drawings describing the highway the major controls for which it is designed. This "design designation" is typified by the following example:

Control of Access = None
ADT 1964 = None
ADT 1984 = 5,000
DHV = 500
D = 60%
T = 5% (assumed)
V = 60 km/h

In addition, the maximum grade and basic lane and shoulder widths may be specified.

Type of Terrain	Minimum Design Speeds (km/h) for Design Volumes of:		
	ADT 0-400	ADT 400-2000	ADT over 2000
Level	60	80	100
Rolling	50	60	80
Mountainous	30	50	60

Table 2-4. Minimum design speeds (rural conditions) (Source: Ref. 1, Table VI-1)

Type of Terrain	Rural Collectors					Design Speed (km/h)			
	30	40	50	60	70	80	90	100	110
	Grades (%)								
Level	7	7	7	7	7	6	6	5	4
Rolling	10	10	9	8	8	7	7	6	5
Mountainous	12	11	10	10	10	9	9	8	6
Design Speed (km/h)									

[a]Maximum grades shown for rural and urban conditions of short lengths, (less than 150 m), on one-way down grades and on low-volume rural collectors may be 2% steeper.

Table 2-5. Maximum grades   (Source: Ref. 1, Table VI-3)

Design Speed (km/h)	Design Traffic Volumes:			
	ADT Under 400	ADT 400-1500	ADT 1500-2000	ADT Over 2000
	Width of Traveled Way (m)[a]			
30	6.0[b]	6.0	6.6	7.2
40	6.0[b]	6.0	6.6	7.2
50	6.0[b]	6.0	6.6	7.2
60	6.0[b]	6.6	6.6	7.2
70	6.0	6.6	6.6	7.2
80	6.0	6.6	6.6	7.2
90	6.6	6.6	7.2	7.2
100	6.6	6.6	7.2	7.2
110	6.6	6.6	7.2	7.2
	Width of Graded Shoulder - Each Side (m)			
All Speeds	0.6	1.5[c]	1.8	2.4

[a] Where the width of the traveled way is shown to be 7.2 m, the width of traveled way may remain at 6.6 m on reconstructed highways where alinement and safety records are satisfactory.

[b] 5.4 minimum for ADT under 250.

[c] May be adjusted to achieve a minimum roadway width of 9.0 m for design speeds of 50 km/h or less.

See text for roadside barrier and offtracking considerations.

Table 2-6. Minimum width of traveled way and graded shoulders (Source: Ref. 1, Table VI-4)

ELEMENTS OF DESIGN

Having established the major design controls, the next step is to relate them to each of the major elements of the highway design. The design of these elements, based upon the fundamental determinants of driver and vehicle characteristics and environmental conditions, includes primarily horizontal and vertical alignment and other factors such as drainage and landscaping.

Stopping Sight Distance -- Stopping sight distance for a given design speed is the minimum distance that a vehicle moving at the corresponding running speed will require to come to a safe halt. It is the sum of the distances traveled during the driver's brake reaction time and during the braking of the vehicle to a stop on a wet pavement. Stopping sight distances for various design speeds are summarized in Table 2-7.

Passing Sight Distance -- Passing sight distance is particularly important when considering safety and alignment. As described by AASHTO, "design passing sight distance is the minimum distance required to safely make a normal passing maneuver on 2-lane highways at passing speeds representative of nearly all drivers, commensurate with design speed. Passing sight distance on 2-lane highways should be provided over as high a proportion of the highway length as feasible. This proportion should be greater on highways with high volumes than on those with low volumes." See Figure 2-2 for passing sight distances for various design speeds.

Horizontal Alignment -- Based upon the selected design speed and the allowable (as specified by state or other jurisdiction) superelevation rate, the minimum radius of curve, or corresponding maximum degree of curvature, may be specified. Table 2-8 indicates the key relationships between design speed, side friction factor, superelevation, minimum radius of curvature, and maximum degree of curve. Typically, establishment of the minimum radius (or degree of curve) is a basic step required before a realistic route selection can be made. For example, for a design speed of 60 km/h and maximum superelevation of 6%, the minimum radius would be 135 m, as shown in Table 2-8.

Design Speed (km/h)	Assumed Speed for Condition (km/h)	Brake Reaction Time (s)	Brake Reaction Distance (m)	Coefficient of Friction[a] f	Breaking Distance on Level (m)	Stopping Sight Distance for Design (m)
30	30-30	2.5	20.8-20.8	0.40	8.8-8.8	29.6-29.6
40	40-40	2.5	27.8-27.8	0.38	16.6-16.6	44.4-44.4
50	47-50	2.5	32.6-34.7	0.35	24.8-28.1	57.4-62.8
60	55-60	2.5	38.2-41.7	0.33	36.1-42.9	74.3-84.6
70	63-70	2.5	43.7-48.6	0.31	50.4-62.2	94.1-110.8
80	70-80	2.5	48.6-55.5	0.30	64.2-83.9	112.8-139.4
90	77-90	2.5	53.5-62.5	0.30	77.7-106.2	131.2-168.7
100	85-100	2.5	59.0-69.4	0.29	98.0-135.6	157.0-205.0
110	91-110	2.5	63.2-76.4	0.28	116.3-170.0	179.5-246.4
120	98-120	2.5	68.0-83.3	0.28	134.9-202.3	202.9-285.6

[a] Values of coefficient of friction generally approximate curves 9 and 10 (coefficient of friction for wet-PC concrete and wet-plant mixes) shown in Figure III-1A.

Table 2-7. Stopping sight distance (wet pavements)
(Source: Ref. 1, Table III-1)

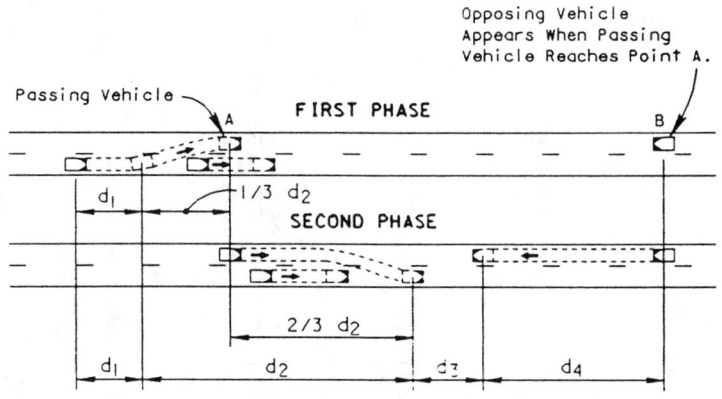

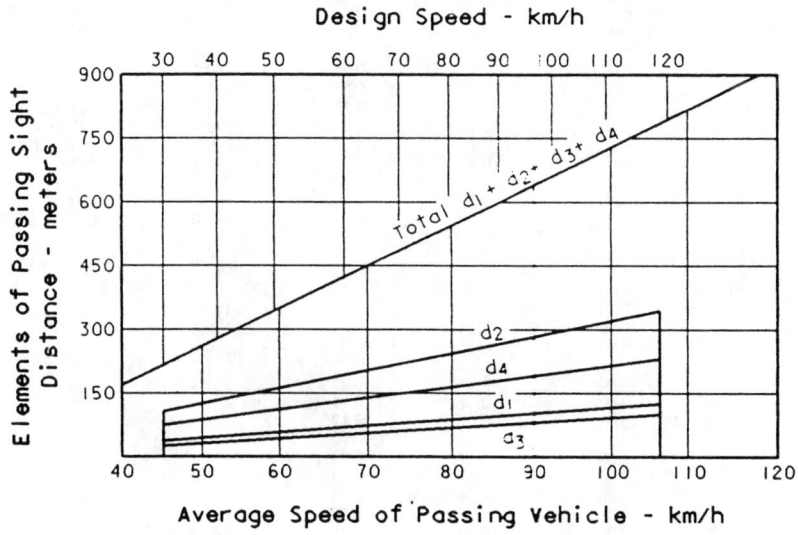

Figure 2-2. Elements of and total passing sight distance -
two-lane highways
(Source: Ref. 1, Figure III-2)

Design Speed (km/h)	Maximum e %	Limiting Values of f	Total (e/100+f)	Calculated Radius (meters)	Rounded Radius (meters)
30	4.00	0.17	0.21	33.7	35
40	4.00	0.17	0.21	60.0	60
50	4.00	0.16	0.20	98.4	100
60	4.00	0.15	0.19	149.2	150
70	4.00	0.14	0.18	214.3	215
80	4.00	0.14	0.18	280.0	280
90	4.00	0.13	0.17	375.2	375
100	4.00	0.12	0.16	492.1	490
110	4.00	0.11	0.15	635.2	635
120	4.00	0.09	0.13	872.2	870
30	6.00	0.17	0.23	30.8	30
40	6.00	0.17	0.23	54.8	55
50	6.00	0.16	0.22	89.5	90
60	6.00	0.15	0.21	135.0	135
70	6.00	0.14	0.20	192.9	195
80	6.00	0.14	0.20	252.0	250
90	6.00	0.13	0.19	335.7	335
100	6.00	0.12	0.18	437.4	435
110	6.00	0.11	0.17	560.4	560
120	6.00	0.09	0.15	755.9	755
30	8.00	0.17	0.25	28.3	30
40	8.00	0.17	0.25	50.4	50
50	8.00	0.16	0.24	82.0	80
60	8.00	0.15	0.23	123.2	125
70	8.00	0.14	0.22	175.4	175
80	8.00	0.14	0.22	229.1	230
90	8.00	0.13	0.21	303.7	305
100	8.00	0.12	0.20	393.7	395
110	8.00	0.11	0.19	501.5	500
120	8.00	0.09	0.17	667.0	665
30	10.00	0.17	0.27	26.2	25
40	10.00	0.17	0.27	46.7	45
50	10.00	0.16	0.26	75.7	75
60	10.00	0.15	0.25	113.4	115
70	10.00	0.14	0.24	160.8	160
80	10.00	0.14	0.24	210.0	210
90	10.00	0.13	0.23	277.3	275
100	10.00	0.12	0.22	357.9	360
110	10.00	0.11	0.21	453.7	455
120	10.00	0.09	0.19	596.8	595
30	12.00	0.17	0.29	24.4	25
40	12.00	0.17	0.29	43.4	45
50	12.00	0.16	0.28	70.3	70
60	12.00	0.15	0.27	105.0	105
70	12.00	0.14	0.26	148.4	150
80	12.00	0.14	0.26	193.8	195
90	12.00	0.13	0.25	255.1	255
100	12.00	0.12	0.24	328.1	330
110	12.00	0.11	0.23	414.2	415
120	12.00	0.09	0.21	539.9	540

NOTE: In recognition of safety considerations, use of e_{max} = 4.00 % should be limited to urban conditions.

Table 2-8. Maximum degree of curve and minimum radius determined for limiting values of e and f, rural highways and high-speed urban streets (Source: Ref. 1, Table III-6)

In addition to the minimum radius, other features of the horizontal alignment must be considered, including:

1. *Transition Curves for Superelevation Runoff.* These should be included in accordance with the selected pavement rotation methods shown in Figure 2-3; the length of runoff and pavement edge slopes of Tables 2-9 and 2-10, respectively; and adherence to accepted practice of the appropriate jurisdiction concerning smoothing of edge profiles.

2. *Pavement Widening on Curves.* This is done in accordance with Table 2-11. Note that pavement widening may not be necessary if the lower design speeds are combined with low degrees of curve.

3. *Sight Distance on Horizontal Curves.* Sight distances must be adequate for the design speed and location of obstructions, as indicated in Figure 2-4.

General Controls for Horizontal Alignment -- In fitting the highway alignment to the terrain, a variety of curves and different lengths of tangent will be appropriate. General controls for establishing the horizontal alignment are listed in Table 2-12 and described in greater detail in Ref. 1.

Vertical Alignment -- The vertical alignment of the highway, as defined by the profile, must conform to necessary design speed and sight distances (described earlier), the terrain, and the characteristics of the vehicles expected to use it.

1. *Design Speed and Terrain.* In general, the design speed increases as the maximum grades decrease, as indicated earlier in the discussion of design speeds.

2. *Critical Grade Lengths.* The critical length of grade is shown in Figure 2-5. The data are based upon observed truck performance. Where critical lengths of upgrade are substantially exceeded, consideration should be given to providing climbing lanes, particularly where truck volumes are high. However, this will rarely be a feature of collector highways because of the lower volumes and speeds associated with this classification.

3. *Lengths of Vertical Curves.* Based upon minimum stopping sight distances for specified design speeds and algebraic differences between connecting grades, the minimum lengths of parabolic crest and sag curves are given in Figures 2-6 and 2-7, respectively.

31

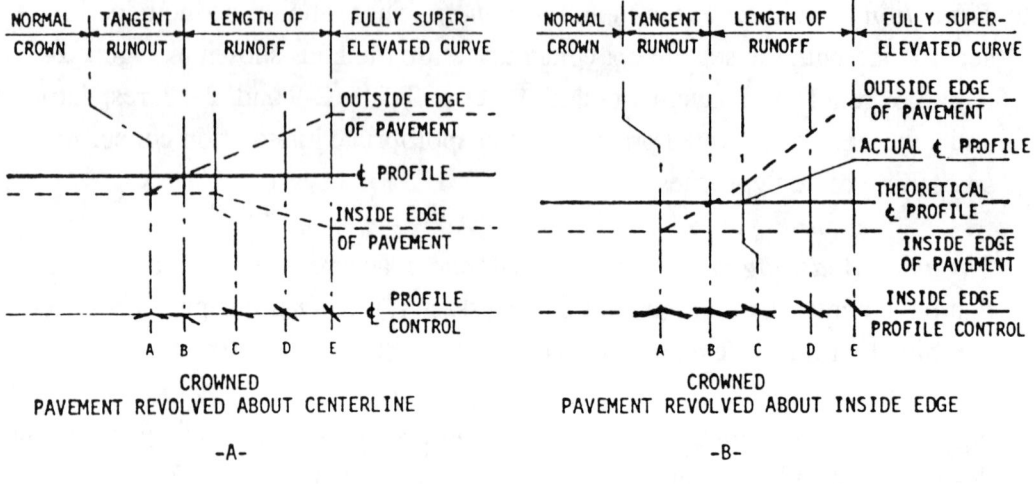

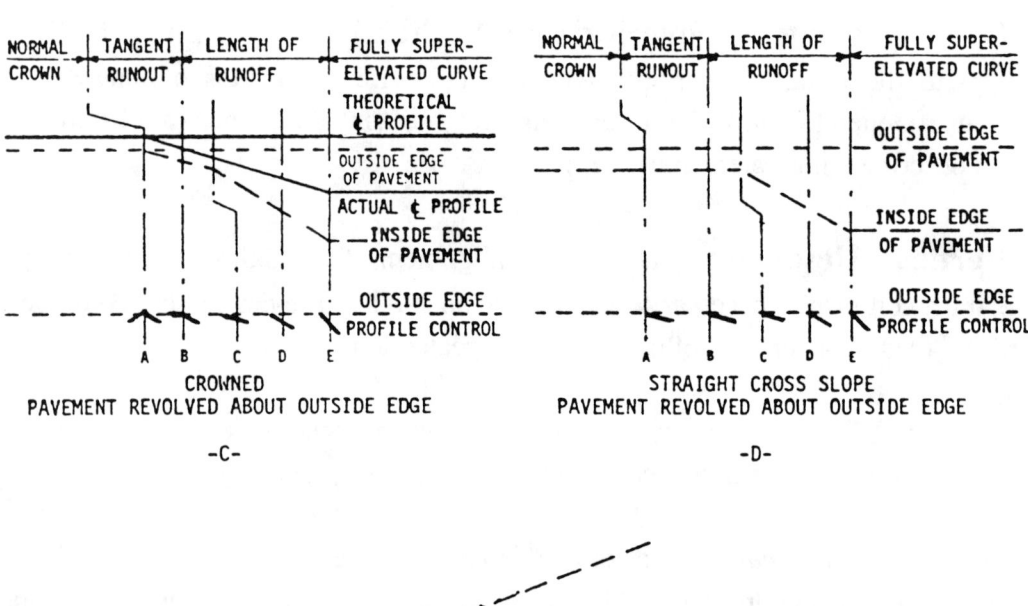

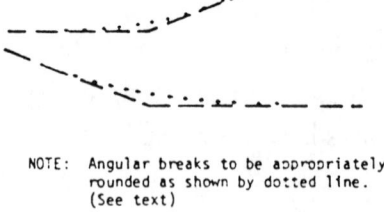

NOTE: Angular breaks to be appropriately rounded as shown by dotted line. (See text)

Figure 2-3. Diagrammatic profiles showing methods of attaining superelevation for curve to the right (Source: Ref. 1, Figure III-16)

Superelevation Rate in %	L—Length of Runoff (m) for Design Speed (km/h) of:									
	30	40	50	60	70	80	90	100	110	120
					3.6 m lanes					
2	20	25	30	35	40	50	55	60	65	70
4	20	25	30	35	40	50	55	60	65	70
6	30	35	35	40	40	50	55	60	65	70
8	40	45	45	50	55	60	60	65	70	75
10	50	55	55	60	65	75	75	80	85	90
12	60	65	65	75	80	90	90	95	105	110
					3.0 m lanes					
2	20	25	30	35	40	50	55	60	65	70
4	20	25	30	35	40	50	55	60	65	70
6	25	30	30	35	40	50	55	60	65	70
8	35	35	40	40	45	50	55	60	65	70
10	40	45	45	50	55	60	65	70	75	75
12	50	55	55	60	65	75	75	80	85	90

Table 2-9. Length required for superelevation runoff-
two-lane pavements (Ref. 1, Table III-14)

Design Speed V_D (km/h)	Maximum Relative Gradients (and Equivalent Maximum Relative Slopes) for Profiles Between the Edge of Two-Lane Traveled Way and the Centerline (%)
30	0.75 (1:133)
40	0.70 (1:143)
50	0.65 (1:150)
60	0.60 (1:167)
70	0.55 (1:182)
80	0.50 (1:200)
90	0.48 (1:210)
100	0.45 (1:222)
110	0.42 (1:238)
120	0.40 (1:250)

Table 2-10. Relationship of design speed to
maximum relative profile gradients
(Source: Ref. 1, Table III-13)

Table 2-11. Calculated and design values for pavement widening on open highway curves (two-lane pavements, one-way or two-way)

Radius of curve (m)	7.2 meters — Design Speed (km/h)								6.6 meters — Design Speed (km/h)								6.0 meters — Design Speed (km/h)						
	50	60	70	80	90	100	110	120	50	60	70	80	90	100	110	120	50	60	70	80	90	100	110
1500	0.0	0.0	0.0	0.0	0.0	0.0	0.0		0.2	0.2	0.2	0.3	0.3	0.4	0.4	0.4	0.3	0.4	0.4	0.4	0.4	0.5	0.6
1000	0.0	0.0	0.1	0.1	0.1	0.1	0.2	0.2	0.3	0.3	0.3	0.4	0.4	0.4	0.5	0.5	0.4	0.4	0.4	0.5	0.5	0.5	0.6
750	0.0	0.0	0.1	0.1	0.1	0.2	0.3	0.3	0.3	0.3	0.3	0.4	0.5	0.5	0.6	0.6	0.6	0.6	0.7	0.7	0.7	0.8	0.8
500	0.2	0.3	0.3	0.4	0.4	0.5	0.5		0.5	0.6	0.6	0.7	0.7	0.8	0.8		0.8	0.9	0.9	1.0	1.0	1.1	1.1
400	0.3	0.3	0.4	0.4	0.5	0.5			0.6	0.6	0.7	0.7	0.8	0.8			0.9	0.9	1.0	1.0	1.1	1.1	
300	0.3	0.4	0.4	0.5	0.5				0.6	0.7	0.7	0.8	0.8				0.9	1.0	1.0	1.1			
250	0.4	0.5	0.5	0.6					0.7	0.8	0.8	0.9					1.0	1.1	1.1	1.2			
200	0.6	0.7	0.8						0.9	1.0	1.1						1.2	1.3	1.3	1.4			
150	0.7	0.8							1.0	1.1	1.1						1.3	1.4					
140	0.7	0.8							1.0	1.1							1.3	1.4					
130	0.7	0.8							1.0	1.1							1.3	1.4					
120	0.7	0.8							1.0	1.1							1.3	1.4					
110	0.7								1.0								1.3						
100	0.8								1.1								1.4						
90	0.8								1.1								1.4						
80	1.0								1.3								1.6						
70	1.1								1.4								1.7						

NOTES: Values less than 0.6 may be disregarded.
2-lane roadways: multiply above values by 1.5.
4-lane roadways: multiply above values by 2.
Where semitrailer volumes are significant up to WB-15/18 increase tabular values of widening by 0.2 for curves with radii of 110 to 175 and by 0.3 for curves with radii of less than 110 . For WB-19's increase values of widening by 0.2 for curves with radii of 225 to 400, by 0.3 for curves with radii 155 to 220, by 0.5 for curves with radii 80 to 150 and by 0.6 for curves with radii less than 80.

For WB-20's increase values of widening by 0.2 for curves with radii 250 to 600, by 0.3 for curves with radii 175 to 245, by 0.4 for curves 155 to 170, by 0.5 for curves with radii 120 to 150, by 0.6 for curves with radii 85 to 115, by 0.8 for curves with radii 70 to 80, and by 0.9 for curves with radii less than 70.

For WB-35's increase values for widening by 0.2 for curves with radii 1000 to 1700, by 0.3 for curves with radii 350 to 995, by 0.4 for curves with radii 305 to 345, by 0.5 for curves with radii 225 to 300, by 0.6 for curves with radii 175 to 220, by 0.7 for curves with radii 155 to 170, by 0.8 for curves with radii 130 to 150, by 1.1 for curves with radii 105 to 125, by 1.2 for curves with radii 85 to 100, by 1.4 for curves with radii 70 to 80, and by 1.5 for curves with radii less than 70.

(Source: Ref. 1, Table III-22)

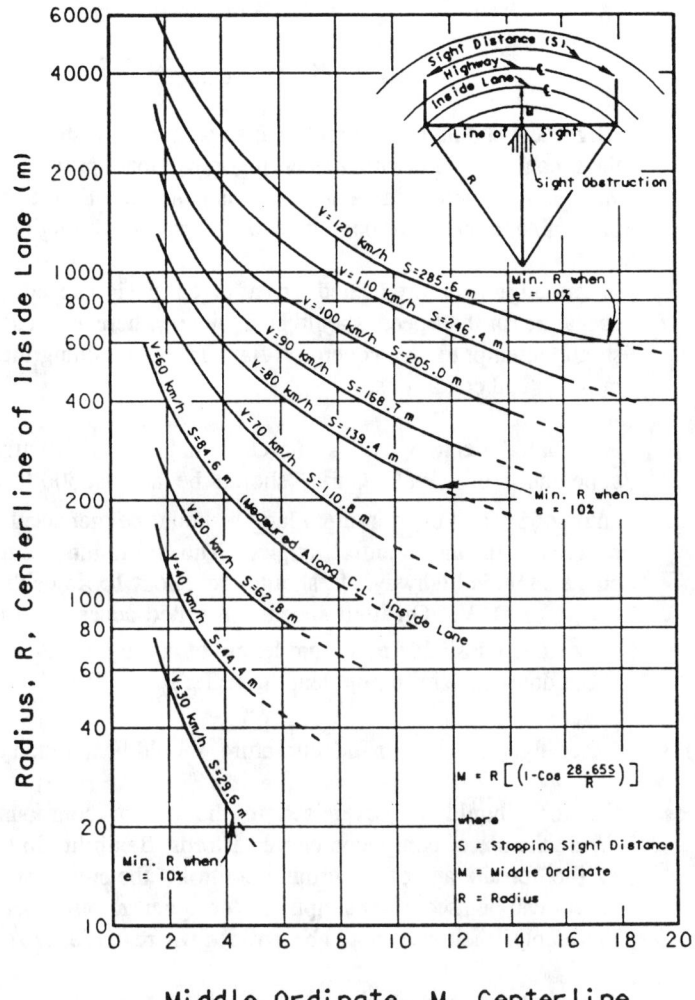

Figure 2-4. Range of upper values-relation between degree of curve and value of middle ordinate necessary to provide stopping sight distance on horizontal curves under open conditions
(Source: Ref. 1, Figure III-24A)

Table 2-12. General controls for horizontal alignment
(Source: Based upon Ref. 1, pp. 224-226)

ITEM	GENERAL CONTROLS
Alignment generally	Alignment should be as directional as possible but should be consistent with the topography and with preserving developed properties and community values. A flowing line that conforms generally to the natural contours is preferable to one with long tangents that slashes through the terrain.
Curve selection	In an alignment predicated on a given design speed, use of maximum curvature for that speed should be avoided wherever possible. The designer should attempt to use generally flat curves, retaining the maximum for the most critical conditions.
Curve length	For small deflection angles, curves should be sufficiently long to avoid the appearance of a kink. Curves should be at least 500 ft long for a central angle of 5°, and the minimum length should be increased 100 ft for each 1° decrease in the central angle. The minimum length of horizontal curve on main highways, L, should be about 15 times the design speed, or $L_{c\ min} = 15V$. On high speed controlled-access facilities that use flat curvature, a desirable minimum length of curve for esthetic reasons would be about double the minimum length, or $L_{c\ des} = 30V$.
Alignment on fills	Other than tangent or flat curvature should be avoided on high, long fills.
Compound curves	Caution should be exercised in the use of compound circular curves. While the use of compound curves affords flexibility in fitting the highway to the terrain and other ground controls, the simplicity with which such curves can be used often tempts the designer to use them without restraint. Preferably, their use should be avoided where curves are sharp.
Reversals	Any abrupt reversal in alignment should be avoided. Such a change makes it difficult for a driver to keep within his own lane. It is also difficult to superelevate both curves adequately, and erratic operation may result. A reversal in alignment can be designed suitably by including a sufficient length of tangent between the two curves for superelevation runoff, or preferably, an equivalent length with spiral curves.
Broken-back curves	The "broken-back" or "flat-back" arrangement of curves (having a short tangent between two curves in the same direction) should be avoided except where very unusual topographical or right-of-way conditions dictate otherwise.
Coordination	To avoid the appearance of inconsistent distortion, the horizontal alignment should be coordinated with the profile design.

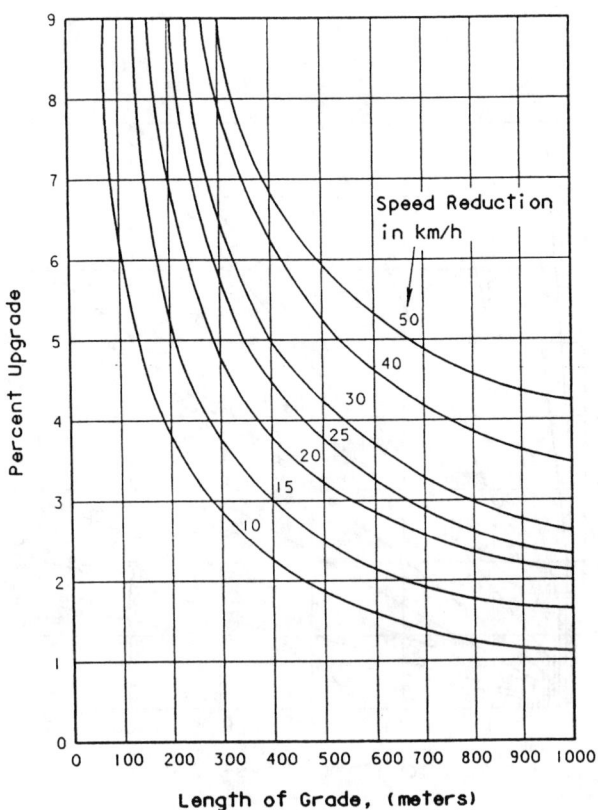

The method of using Figure III-29 to determine critical lengths of grade is demonstrated in the following examples.

Assume that a highway is being designed for 100 km/h and has a fairly level approach to a 4 percent upgrade; direct reading on the heavy line in Figure III-29 shows the critical length to be 280 m. If instead, the design speed were 60 km/h, the initial and minimum tolerable speeds on the grade would be different, but for the same permissible speed reduction the critical length would still be 280 m.

In another instance, the critical length of a 5 percent upgrade approached by a 500 m length of 2 percent upgrade is unknown. Figure III-29 shows that 500 m of 2 percent upgrade results in a speed reduction of about 11 km/h. The chart further shows that the remaining allowable speed reduction of 4 km/h will be made on 100 m of 5 percent upgrade.

Figure 2-5. Critical lengths of grade for design, assumed typical heavy truck of 180 kg/kW, entering speed = 90 km/h (Source: Ref. 1, Figure III-29)

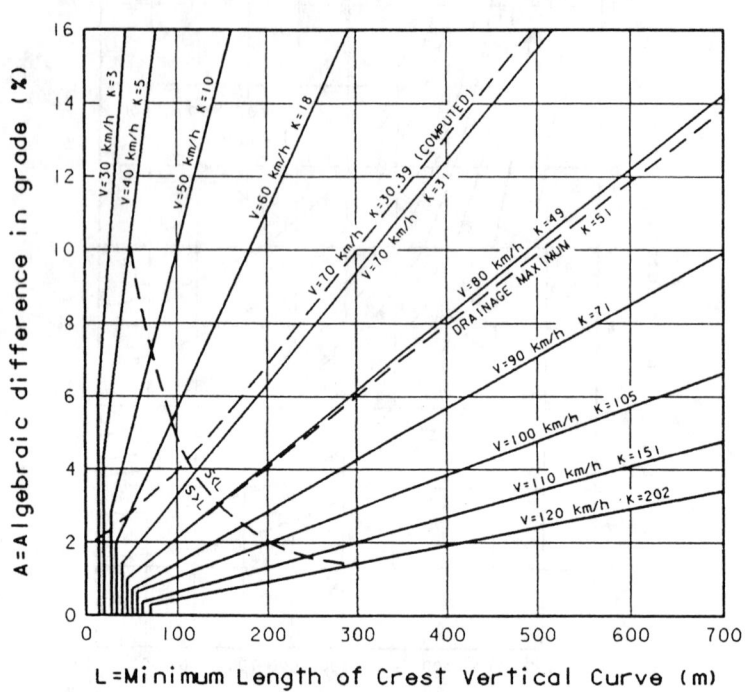

Figure 2-6. Design controls for crest vertical curves, for stopping sight distance - upper range)
(Source: Ref. 1, Figure III-39)

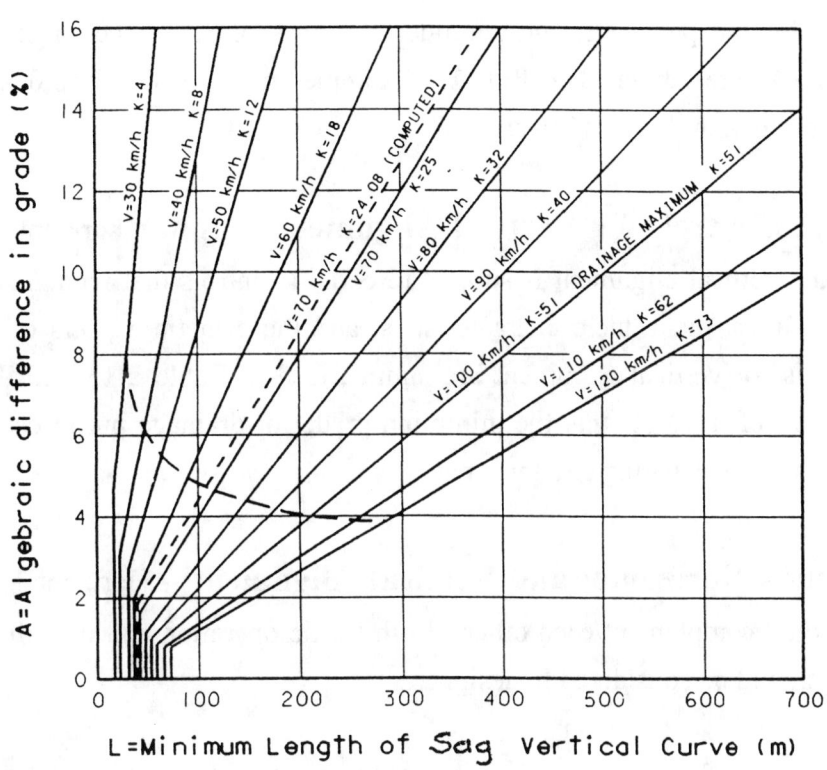

**Figure 2-7. Design controls for sag vertical curves - upper range
(Source: Ref. 1, Figure III-41)**

The criterion for rate of change of grade, K, is such that when K = 51, the grade at a point 15 m from the minute horizontal portion of the curve will be 0.30%. This is considered adequate for drainage purposes, but other values of the grade at this location may be specified. For example, some state highway departments specify a maximum value of K = 40, equivalent to a grade of 0.40% at a point 15 m from the horizontal point of the curve. Note that when K is greater than the specified value, special precautions to provide adequate drainage are usually required. These precautions may include provision of additional catch-basins or flumes, or increased crossfall for the pavement and shoulder. Values of K related to stopping sight distance also are tabulated in Ref. 1. Examples of these for the 70 km/h design speed are shown in Figures 2-6 and 2-7.

General Controls for Vertical Alignment -- As with horizontal alignment, provision of the vertical alignment is subject to certain controls that recognize a driver's task and capabilities, the vehicle characteristics, and the esthetic aspects of the design. General controls for vertical alignment are summarized in Table 2-13 and described in greater detail in Ref. 1. Note that the minimum grade for drainage purposes is generally 0.5%, except under extenuating circumstances.

Combined Horizontal and Vertical Alignment -- Horizontal and vertical alignments should complement each other. Both traffic operation and overall appearance of the facility should be considered in design.

As stated in Ref. 1, vertical curvature superimposed upon the horizontal or vice versa generally results in a pleasing facility. Important exceptions to these alignment combinations (Ref. 1) are particularly relevant to the project outlined later, and include:

1. A series of vertical "humps" not in combination with horizontal curvature may result in drivers' sight distance problems.

2. Sharp horizontal curvature should not be introduced at or near the top of a pronounced crest vertical curve.

3. Only flat horizontal curvature should be introduced at or near the low point of a pronounced sag vertical curve.

Table 2-13. (General controls for vertical alignment
(Source: Ref. 1, pp 295-296)

ITEM	GENERAL CONTROLS
Gradeline generally	A smooth gradeline with gradual changes, as consistent with the type of highways, roads, or streets and the character of terrain, should be strived for in preference to a line with numerous breaks and short lengths of grade.
"Rolling" gradelines	The "roller-coaster" or the "hidden-dip" type of profile should be avoided. Such profiles generally occur on relatively straight horizontal alignment where the roadway profile closely follows a rolling natural ground line. Examples of these undesirable profiles are evident on many older roads and streets. They are unpleasant esthetically and more difficult to drive.
Undulating gradelines	Undulating gradelines, involving substantial lengths of momentum grades, should be appraised for their effect on traffic operation. Such profiles permit heavy trucks to operate at higher overall speeds than is possible when an upgrade is not preceded by a downgrade, but may encourage excessive speeds of trucks with attendant conflicts with other traffic.
Broken-back gradelines	A broken-back gradeline (two vertical curves in the same direction separated by a short section of tangent grade) generally should be avoided, particularly in sags where the full view of both vertical curves is not pleasing.
Sustained grade breaks	On long grades it may be preferable to place the steepest grades at the bottom and lighten the grades near the top of the ascent or to break the sustained grade by short intervals of lighter grade instead of a uniform sustained grade that might be only slightly below the allowable maximum.
Intersections	Where intersections-at-grade occur on roadway sections with moderate to steep grades, it is desirable to reduce the gradient through the insection. Such a profile change is beneficial for all vehicles making turns and serves to reduce the potential hazards.
Sag curves in cuts	Sag vertical curves should be avoided in cuts unless adequate drainage can be provided.

4. On two-lane highways, the need for safe passing sections at frequent intervals and for an appreciable percentage of the highway length may affect the extent of coordination between the horizontal and vertical alignments.

5. Horizontal curvature and profile should be made as flat as feasible at highway intersections.

Signs and Markings -- So that adequate warning is given to drivers about the geometric and other features of the highway, the provision of appropriate signs, markings, and other traffic control devices must be considered. Speed-control signs and restricted passing locations are particularly important. The <u>Manual on Uniform Traffic Control Devices</u> (MUTCD, Ref. 3) should be consulted for details of the location and design of traffic control devices.

Other Elements Affecting Geometric Design -- Depending upon the nature and location of the proposed highway, other elements in addition to drainage systems, erosion control, and landscaping (discussed later) may have to be considered. These elements include:

- roadside turnouts and rest areas
- driveways and roadside control
- lighting
- utilities.

Because the emphasis in this book is on the choice and selection of route and preliminary rather than final design, the above four elements are not discussed further, and the reader should consult Ref. 1 and other sources of information for further details.

CROSS SECTION ELEMENTS AND ROADSIDE DESIGN

In addition to the horizontal and vertical alignment, the major cross-section elements and associated roadside design for the highway must be considered, particularly in cities, mountainous terrain, or other locations where the amount of available space for the highway and its appurtenances is limited.

Typical Cross Section Elements -- The cross section elements that must be considered and shown on the drawings for a two-lane rural highway are the pavement surface, shoulders, side and back slopes, ditches, curbs, and right-of-way locations. Typical policy as described in Ref. 1 with regard to cross-section elements is summarized as follows:

1. *Surface Type and Cross Slope* -- For high-type pavements, a cross slope of 1.5% to 2% in tangent sections, increasing to 3% for intermediate-type pavements, is suggested in Ref. 1.

2. *Lane Width* -- A lane width of at least 3.6 m is preferable. This was indicated earlier in the discussion about design speed selection.

3. *Curbs* -- Placement of curbs usually closely follows the guidelines of each state's highway agency and/or in accordance with Ref. 5. Barrier curbs are typically placed so as to protect pedestrians and to minimize collisions between vehicles and structures such as bridge piers and railings. Mountable curbs are usually used to delineate changes in the surface elevations where the effects of a vehicle's straying from the traveled way would not be severe. It should be noted that curbs are not usually a feature of rural highways unless justified by the need for drainage or for sidewalks to accommodate pedestrians.

4. *Shoulders* -- A graded shoulder at least 3.6 m wide is desirable on heavily traveled, high-speed highways, but this may be reduced to 2.4 m as shown in Table 2-6, for a rural collector highway similar to that in the design project described later. Guardrails should be set at least 0.6 m outside the shoulder's outside edge. Shoulder cross slopes should provide adequate drainage and recommended cross slopes are 2% to 6% for bituminous and concrete surfaces, 4% to 6% for gravel, and about 8% for turf surfaces. Shoulders should be stable enough to support occasional vehicle loads under any weather conditions and should ideally contrast with the through traffic lanes in both color and texture. Typical cross sections are shown in Figure 2-8.

5. *Sidewalks* -- In general, wherever the roadside and land development conditions are such that pedestrians regularly move along a rural main or high-speed highway, a sidewalk or path area well removed from the traveled way should be provided.

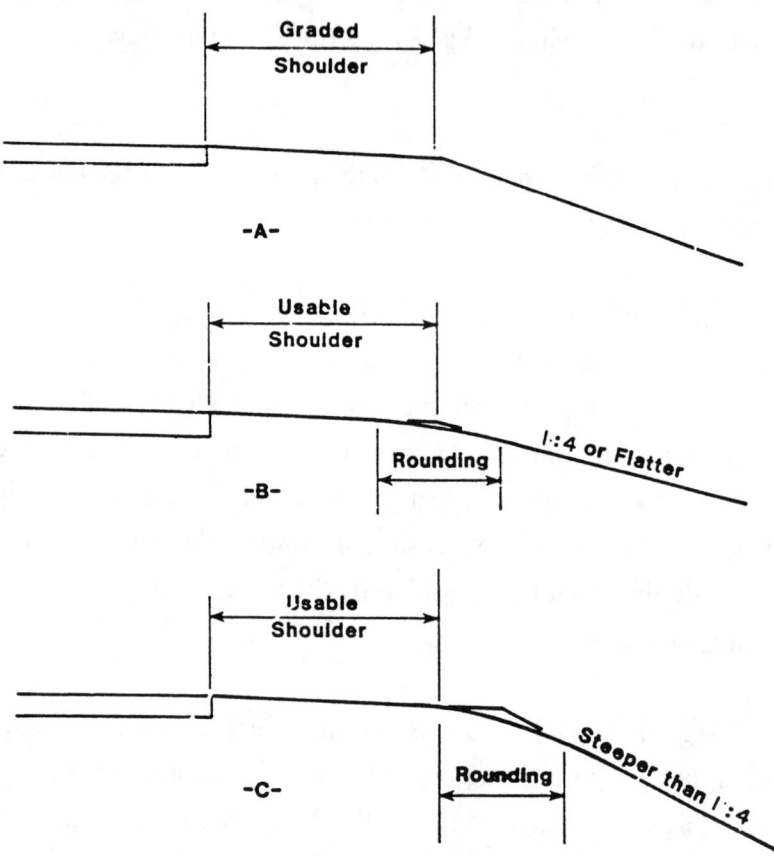

**Figure 2-8. Graded and usable shoulders
(Source: Ref. 1, Figure IV-2)**

6. *Guardrails and Guideposts* -- Generally, guardrails should be provided adjacent to fixed objects along the pavement edge where fills have side slopes steeper than 1 in 4 or are higher than 2.4 m, and where other conditions may be considered hazardous. Many state highway agencies have developed their own guidelines for guardrail use. At less hazardous but similar locations, guideposts are desirable to outline the roadway.

7. *Drainage Channels and Side Slopes* -- To promote safety, appearance, and economy in maintenance, the use of reasonably flat side slopes, broad drainage channels, and extensive rounding of the cross-section slope lines is recommended. Typical cross sections are shown in Figure 2-9.

Roadside Design -- The design of safe roadside areas includes consideration of many factors, including prevailing traffic characteristics and location of "non-traversable" features along the roadside. These features include utility poles, bodies of water, steep slopes, and objects along the roadside that would impede an errant vehicle movement and/or cause serious injury to passengers. The location and design of guardrail or other safety barriers (which in themselves may constitute obstructions) is a part of the design process as well and is described in several publications, particularly Refs. 4 and 5.

A full description of roadside design features is beyond the scope of this book, but the use of two concepts -- the clear zone distance and the warrants for roadside barriers, such as guardrails -- are briefly illustrated below. Some important definitions, based upon Ref. 5, are as follows:

1. Embankment or fill slopes parallel to the flow of traffic may be defined as recoverable, non-recoverable, traversable, or critical.

2. Recoverable slopes are all embankment slopes 4:1 or flatter. If such slopes are relatively smooth and traversable, the suggested clear zone distance (the width of an area parallel to the highway where no impediments to recoverable movements of an errant vehicle occur) can be read from Figure 2-10. Motorists who encroach on recoverable slopes can generally stop their vehicles or slow them enough to return to the roadway safely.

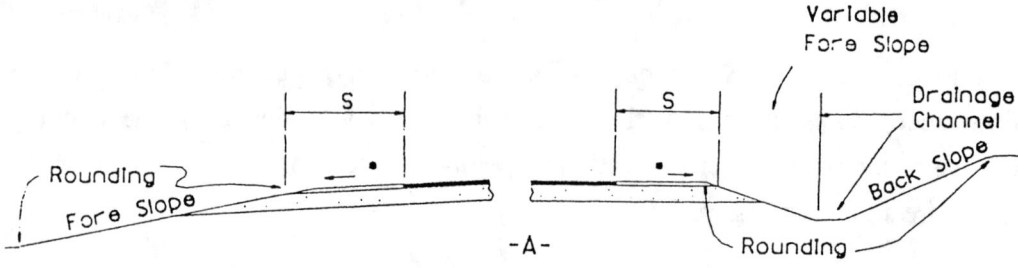

-A-

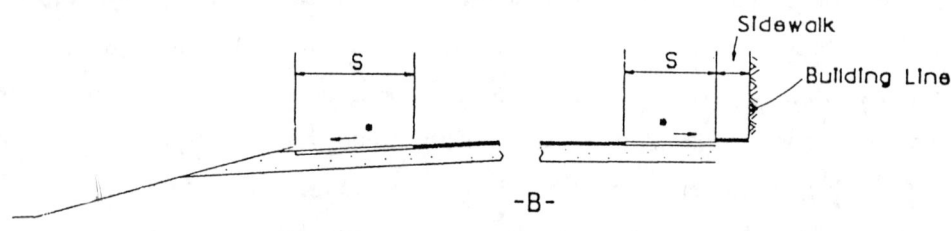

-B-

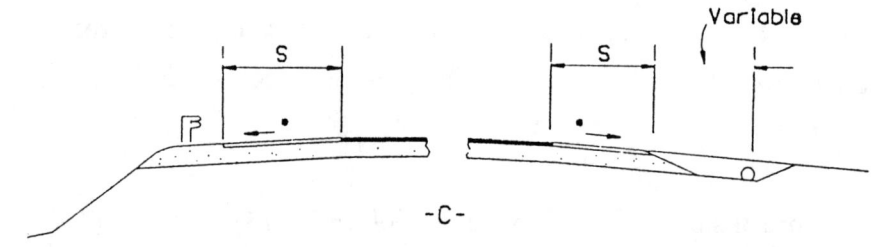

-C-

NOTE:

S = Usable Shoulder

• = Rate of Slope 2 to 3 Per Cent

Figure 2-9. Typical cross section, normal crown.
(Source: Ref. 1, Figure IV-6)

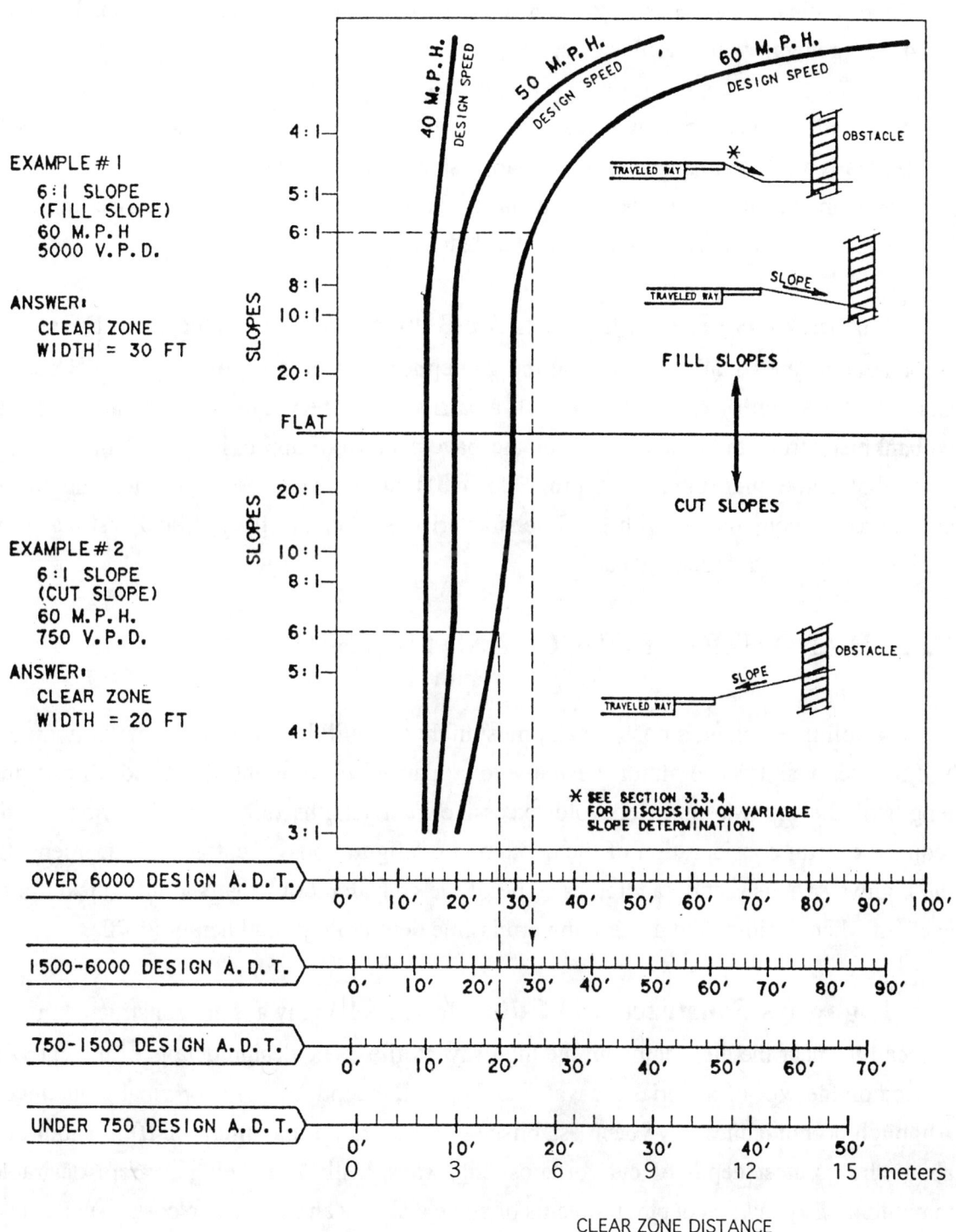

EXAMPLE # 1
  6:1 SLOPE
  (FILL SLOPE)
  60 M.P.H
  5000 V.P.D.

ANSWER:
  CLEAR ZONE
  WIDTH = 30 FT

EXAMPLE # 2
  6:1 SLOPE
  (CUT SLOPE)
  60 M.P.H.
  750 V.P.D.

ANSWER:
  CLEAR ZONE
  WIDTH = 20 FT

OBSTACLE

TRAVELED WAY

FILL SLOPES

CUT SLOPES

TRAVELED WAY   SLOPE   OBSTACLE

* SEE SECTION 3.3.4
FOR DISCUSSION ON VARIABLE
SLOPE DETERMINATION.

OVER 6000 DESIGN A.D.T.

1500-6000 DESIGN A.D.T.

750-1500 DESIGN A.D.T.

UNDER 750 DESIGN A.D.T.

CLEAR ZONE DISTANCE

Figure 2-10. Clear zone distance curves
(Source: Ref. 5, Figure 3.1)

47

3. A non-recoverable slope is defined as one that is traversable, but one on which most motorists will be unable to stop or return to the roadway easily. Vehicles on such slopes typically can be expected to reach the bottom. Embankments between 3:1 and 4: A 1 generally fall into this category.

4. A critical slope is one on which a vehicle is likely to overturn. Slopes steeper than 3:1 generally fall into this category. If a slope steeper than 3:1 begins closer to the traveled way than the suggested clear zone distance for that specific roadway, a barrier might be warranted if the slope cannot readily be flattened.

Comparative Risk Warrants for Barriers -- Warrants for use of barriers on embankments are usually based upon the guidelines presented in Figure 2-11. Here, for example, for an embankment slope of 2:1, a barrier would be warranted if the height of the embankment from the outside edge of the pavement were approximately 2 m or more. Note that some judgement is appropriate and that the need for a barrier should be considered in conjunction with traffic characteristics, accident history, and estimates of the clear zone width described earlier.

DEPTH AND HEIGHT OF CUT AND FILL SECTIONS

In rolling or mountainous terrain, most highways will be constructed partially on cut and fill sections. In general, it is desirable to "balance" the amount of cut and fill over the length of the highway and to avoid excessive haul lengths during construction. This requires careful consideration of the depth and configurations of cut and fill segments of the highway. This section addresses some of the considerations that should be taken into account when deciding the maximum permissible depth of cuts and height of fills.

Highways Constructed in Cuts -- In general, highways are constructed in cuts at locations near the high points in the highway profile. The design details of the cut will depend on the type of material excavated and upon the depth from the original grade level. Although no guidelines have been established, in general, a maximum of 10 m should be about the greatest depth of cut for most highways, with 7 m being a more desirable maximum. Beyond this depth, problems of slope stability and excessive costs usually tend to outweigh the benefits derived from decreases in grades and horizontal alignments, unless the highway accommodates high traffic volumes. Deeper cuts may be necessary but the highway alignment should be investigated very carefully for alternatives before such action is taken.

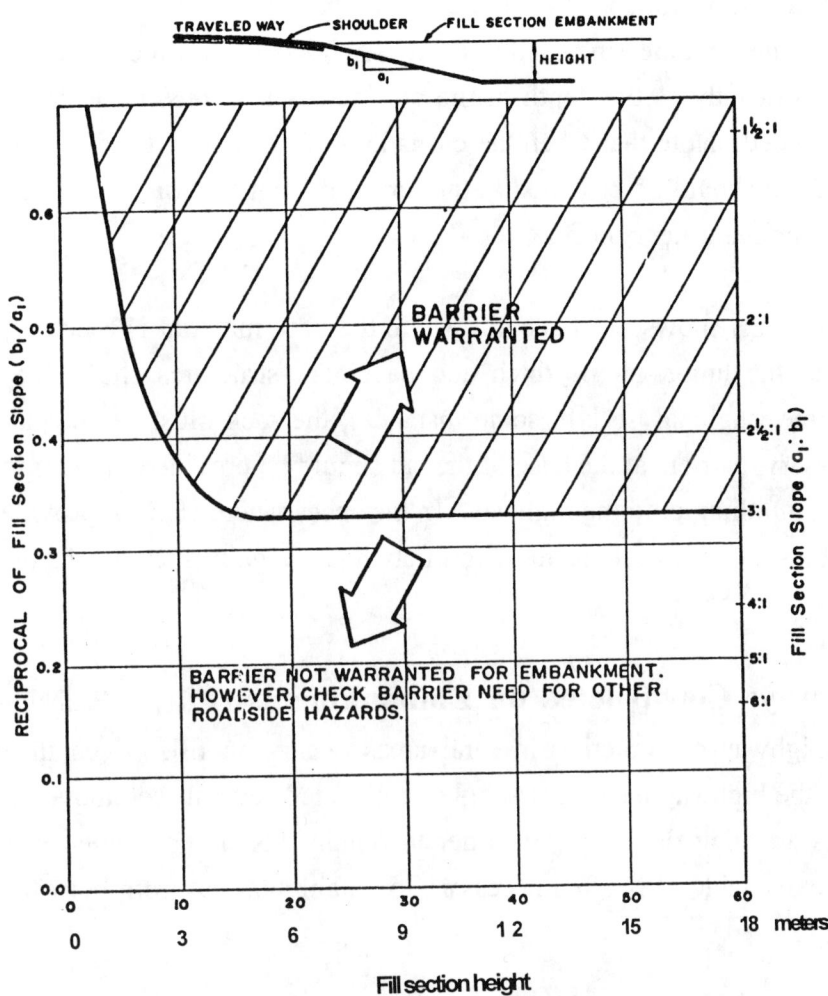

Figure 2-11. Comparative risk warrants for embankments
(Source: Ref. 5, Figure 5.1)

49

An example of selected characteristics of highways constructed in cuts is shown in Figure 2-12. Figure 2-12A shows a highway in a cut where the ground material is stable enough to be formed into self-supporting slopes. In general, slopes should conform to the values described and should be consistent with the provision of guardrail and other safety measures described earlier. The design of all slopes should be based upon detailed engineering soils analysis. In addition, to assist in slope stabilization, fabric and vegetation may be included. Selection of the appropriate materials, again, should be subject to detailed engineering analysis. It can be seen that the volume of excavation increases significantly as the depth of the cut increases. Thus, excessive depths of cuts should be avoided. Note that when the cut is located with an extensive slope above it, an interceptor ditch should be provided at the top of the cut section to reduce erosion of the cut slope by surface water runoff.

Figure 2-12B shows a cut section where the base material is rock. Again, the rock slope beyond the limits of the ditch and pavement structures should be the result of detailed engineering analysis. In some instances, the rock must be adequately stabilized and, if necessary, barriers at the base of the cut should be provided in order to prevent rock fragments from falling onto the highway. In instances where cuts are provided in rock, the expense of blasting quickly becomes extremely high. Again, such depths of cut should be minimized.

Highways Constructed on Embankments (Fill) -- Probably the greatest amount of highway construction in rural areas occurs on fill. Even in extremely flat topography, the highway pavement should be elevated several feet above the surrounding ground, thus assisting drainage. In order to obtain this fill, particularly where few cut segments are available, it is often necessary to obtain material from borrow pits along the highway route.

As with the cut section, it is usually desirable to keep the height of the fill section to 10 m or less, with 7 m being a preferred maximum. Above this height, depending upon the topography, the classification of the highway, and the affected land uses, it may be more economical to construct a bridge. This may be particularly true where the highway passes through rocky terrain or where the fill section is on a marsh, swamp, or any other location where unstable ground conditions occur.

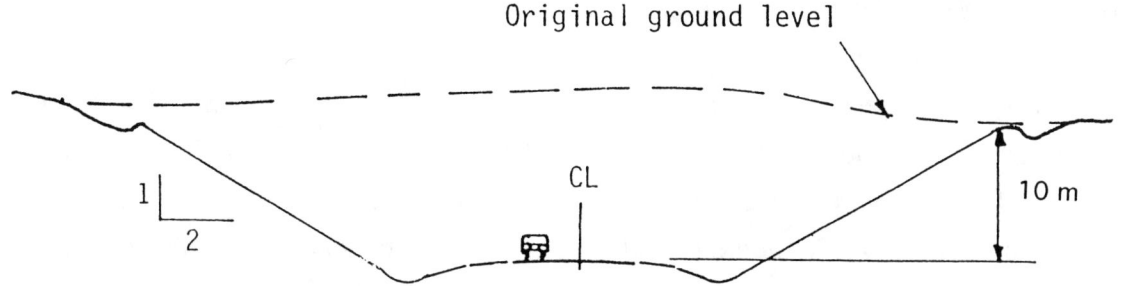

A) Typical conditions with approximately 2:1 slope

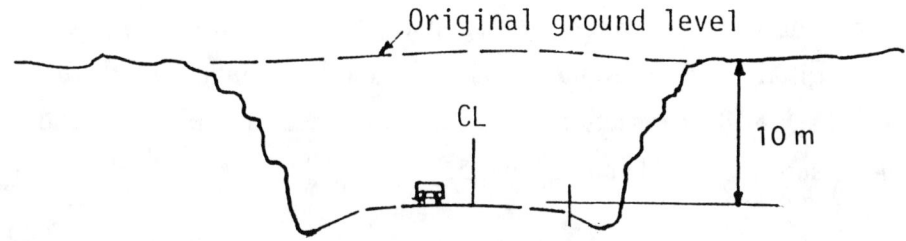

B) Excavation conditions in rock

Note: Each of the above examples shows a 10 m deep cut.
This is a considerable depth for a two-lane
highway and will be close to the maximum practicable depth in most cases.

Figure 2-12. Highways in cut sections

Slopes for cuts and fills should be determined in accordance with the guidelines discussed earlier under Cross Section Elements, including the provision of guardrails or other safety devices. Usually, the side slopes should be no steeper than 1 in 2 for regular fill material from soil stability, surface vegetation, and maintenance considerations. This slope may be increased somewhat if rock or other more stable material is available for the purpose or if limited space or environmental factors dictate. **Again, the exact slope should be the result of detailed engineering analysis.**

Drainage of fill sections is particularly important to ensure that rainwater from the highway surface does not erode the slopes of the fill, resulting in possible eventual subsidence of the highway surface itself. Thus, curbing at the top of the fill, transverse flumes, and substantial ditches at the foot of the slope section may be major features of the drainage systems.

INTERSECTIONS AT GRADE

Many combinations of curb radii and tapers may be used to ensure that vehicles of different sizes can safely and conveniently negotiate at-grade intersections. One of the most basic methods is the use of tabulated values of radii and taper length for specific design vehicles and angles of intersection. These are presented in Table 2-14, and may be used in the projects described in Chapters 4 and 5.

ECONOMIC ANALYSIS

It was mentioned earlier and shown in Figure 1 that following an initial geometric design, cost estimate, and economic analysis, the design may have to be re-examined. A higher estimated cost than expected, or a possibly less costly route becoming apparent could be possible reasons for the re-examination. The economic evaluation of a proposed highway -- beyond a basic construction cost estimate -- may then be appropriate. The evaluation may be conducted by means of (a) benefit-cost ratio, (b) net present value, (c) comparison of annual costs, and (c) internal rate of return.

Angle of Turn (degrees)	Design Vehicle	Simple Curve Radius (m)	Simple Curve Radius with Taper		
			Radius (m)	Offset (m)	Taper (m:m)
30	P	18	—	—	—
	SU	30	—	—	—
	WB-12	45	—	—	—
	WB-15	60	—	—	—
	WB-19	110	67	1.0	15:1
	WB-20	116	67	1.0	15:1
	WB-29	77	37	1.0	15:1
	WB-35	145	77	1.1	20:1
45	P	15	—	—	—
	SU	23	—	—	—
	WB-12	36	—	—	—
	WB-15	53	36	0.6	15:1
	WB-19	70	43	1.2	15:1
	WB-20	76	43	1.3	15:1
	WB-29	61	35	0.8	15:1
	WB-35	—	61	1.3	20:1
60	P	12	—	—	—
	SU	18	—	—	—
	WB-12	28	—	—	—
	WB-15	45	29	1.0	15:1
	WB-19	50	43	1.2	15:1
	WB-20	60	43	1.3	15:1
	WB-29	46	29	0.8	15:1
	WB-35	—	54	1.3	20:1
75	P	11	8	0.6	10:1
	SU	17	14	0.6	10:1
	WB-12	—	18	0.6	15:1
	WB-15	—	20	1.0	15:1
	WB-19	—	43	1.2	20:1
	WB-20	—	43	1.3	20:1
	WB-29	—	26	1.0	15:1
	WB-35	—	42	1.7	20:1
90	P	9	6	0.8	10:1
	SU	15	12	0.6	10:1
	WB-12	—	14	1.2	10:1
	WB-15	—	18	1.2	15:1
	WB-19	—	36	1.2	30:1
	WB-20	—	37	1.3	30:1
	WB-29	—	25	0.8	15:1
	WB-35	—	35	0.9	15:1
105	P	—	6	0.8	8:1
	SU	—	11	1.0	10:1
	WB-12	—	12	1.2	10:1
	WB-15	—	17	1.2	15:1
	WB-19	—	35	1.0	30:1
	WB-20	—	35	1.0	30:1
	WB-29	—	22	1.0	15:1
	WB-35	—	28	2.8	15:1
120	P	—	6	0.6	10:1
	SU	—	9	1.0	10:1
	WB-12	—	11	1.5	8:1
	WB-15	—	14	1.2	15:1
	WB-19	—	30	1.5	25:1
	WB-20	—	31	1.6	25:1
	WB-29	—	20	1.1	15:1
	WB-35	—	26	2.8	15:1
135	P	—	6	0.5	15:1
	SU	—	9	1.2	8:1
	WB-12	—	9	2.5	6:1
	WB-15	—	12	2.0	10:1
	WB-19	—	24	1.5	20:1
	WB-20	—	25	1.6	20:1
	WB-29	—	19	1.7	15:1
	WB-35	—	25	2.6	15:1

Table 2-14. Minimum edge-of-pavement designs for turns at intersections (Source: Ref. 1, Table IX-1)

For these methods it is required to know or estimate the following inputs (Ref. 7):

1. Construction or initial investment cost
2. Maintenance cost, usually expressed as an annual amount
3. User cost (fuel, oil, tires, repairs, maintenance, purchase price, and accidents) in using the highway
4. economic analysis period
5. traffic volume
6. interest rate
7. User time and accident costs if included as economic costs

From a designer's point of view in determining the desired route, the ways in which the cost can be reduced relate primarily to items 1 through 3 above. The ways in which this can be done (bearing in mind the design designation and controls, and desirable practice) for each of these three items is as follows:

- In order to minimize construction and maintenance costs, the highway should be made as short as possible and amounts of cut and fill should be minimized.

- In order to minimize user costs (i.e., the cost of operating the vehicle) grades should be as flat as possible and horizontal alignment should avoid the use of sharp curves.

Clearly, the balance between these features can only result from considerable trial and error, and a change in one feature will invariably result in changes to the others. In order to provide some guidance and to estimate the relevant user costs, Figure 2-13 provides several graphs showing the relationship between design speed and operating costs as they are affected by grade and horizontal curvature, as well as an example of their use. The graphs will be used for the economic cost analysis illustrated in the project of Chapter 4.

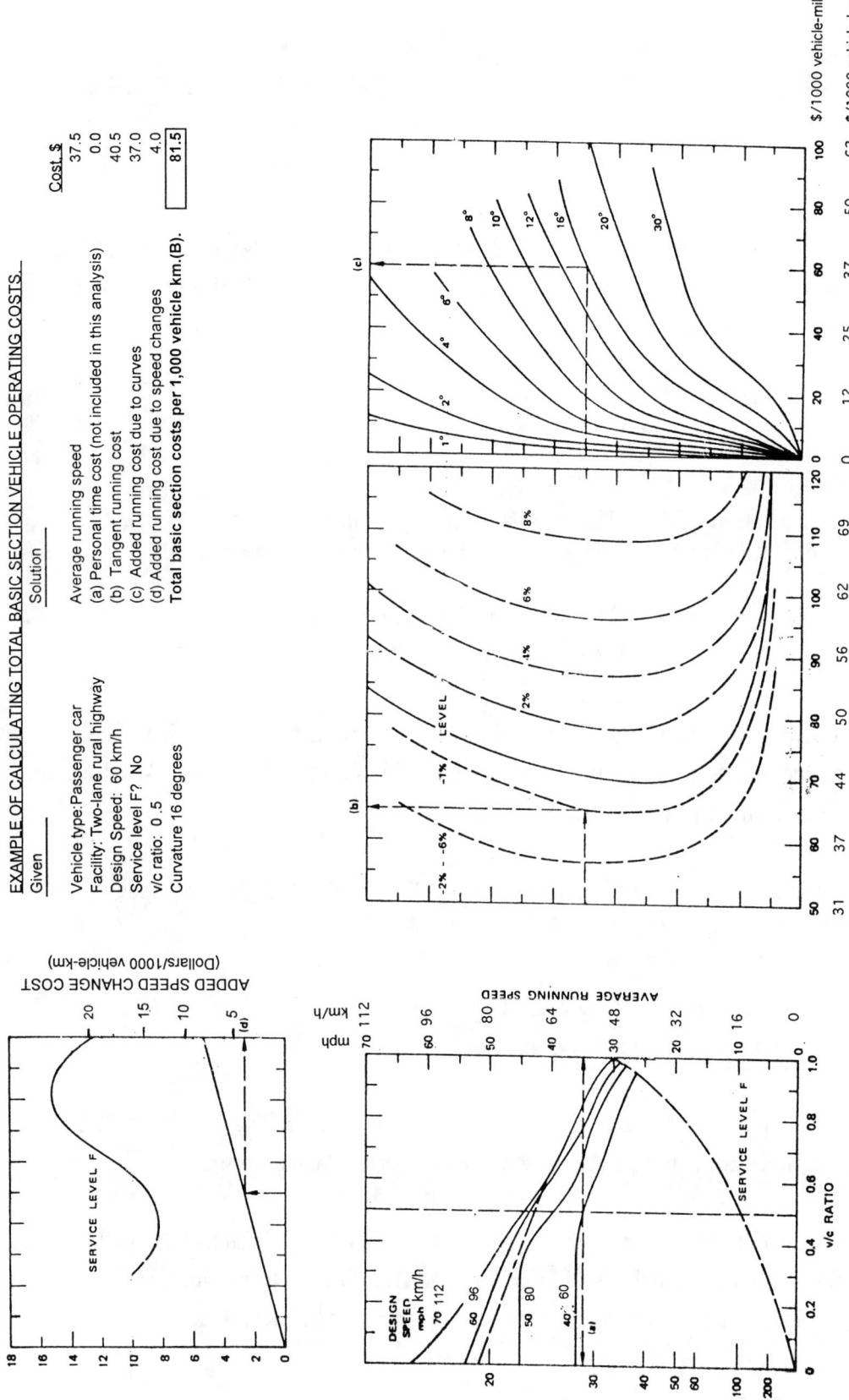

EXAMPLE OF CALCULATING TOTAL BASIC SECTION VEHICLE OPERATING COSTS.

Given	Solution	Cost $
	Average running speed	
Vehicle type: Passenger car	(a) Personal time cost (not included in this analysis)	37.5
Facility: Two-lane rural highway		0.0
Design Speed: 60 km/h	(b) Tangent running cost	40.5
Service level F? No	(c) Added running cost due to curves	37.0
v/c ratio: 0.5	(d) Added running cost due to speed changes	4.0
Curvature 16 degrees	Total basic section costs per 1,000 vehicle km.(B).	81.5

Figure 2-13. Basic section costs for passenger cars on two-lane highways
(Source: Based upon Figure 9, Ref. 7)

55

ENVIRONMENTAL REQUIREMENTS

As shown in Figure 1 and mentioned in Chapter 1, addressing environmental concerns is a major feature of any proposed highway project. Elaborating on the general areas of concern mentioned in Chapter 1 and refering specifically to items 08 and 09 in Table 1-1, state and federal laws and requirements must be addressed. As well as air and noise pollution effects, the use of energy, visual intrusion, examples of federal and state laws required in Massachusetts, <u>particularly those affecting the detailed selection of the proposed route</u>, are listed in Table 2-15. Examples of the reporting are shown in Chapter 3 in relation to the route design process.

REFERENCES

1. American Association of State Highway and Transportation Officials. (1994). "A Policy on Geometric Design of Highways and Streets." Washington, D.C.

2. American Association of State Highway and Transportation Officials. (1965). "A Policy on Geometric Design of Rural Highways." Washington, D.C.

3. Federal Highway Administration, National Advisory Committee on Uniform Traffic Control Devices. (1988). "Manual on Uniform Traffic Control Devices." Washington DC: U.S. Government Printing Office.

4. Transportation Research Board, Special Report 209. (1994). "Highway Capacity Manual." Washington, D.C.

5. American Association of State Highway and Transportation Officials. (1990). "Roadside Design Guide." Washington, D.C.

6. American Association of State Highway and Transportation Officials. (1977). "Guide for Selecting, Locating, and Designing Traffic Barriers." Washington, D.C.

7. American Association of State Highway and Transportation Officials. (1977). "A Manual on User Benefit Analysis of Highway and Bus Transit Improvements." Washington, D.C. (This book is sometimes referred to as the "Red Book").

09 DETERMINE OTHER APPLICABLE FEDERAL, STATE, AND LOCAL ENVIRONMENTAL LAWS AND REQUIREMENTS

The designer, in coordination with the Environmental Section will be responsible for identifying and complying with all other applicable Federal, State and local environmental laws and requirements. A listing of the most common laws and requirements follows. The designer should consult with the Environmental Section to determine if other less common laws or requirements are applicable.

Federal Laws and Requirements

Section 4(f) of 1966 U.S. D.O.T Act — FHWA approval is needed for any Federal-aid highway project using land from a significant publicly owned park, recreation area, historic property or wildlife and waterfowl refuge. A historic property may be a bridge structure, site, or district. An individual or programmatic Section 4(f) evaluation document must be prepared. There must also be coordination with the public official having jurisdiction over the Section 4(f) property. Additional details about Section 4(f) process are discussed with the MDPW Environmental Section.

Section 404 of 1972 Clean Water Act (33USC1344) — A permit is required from the U.S. Army Corps of Engineers for highway projects involving discharge of dredged or fill material into waters of the United States. Jurisdiction under this law extends to lakes, rivers, streams, wetlands, and mudflats. There are four classes of permits issued: nationwide, regional, general, and individual.

An individual permit also involves consultation with the U.S. Environmental Protection Agency and the U.S. Fish and Wildlife Service. A water quality certification and a coastal zone consistency statement (if applicable) is needed before the U.S. Corps of Engineers will issue the Section 404 permit. Additional details about the Section 404 permit process are discussed with the MDPW Environmental Section.

Section 401 of 1972 Clean Water Act — A water quality certification is required from the Massachusetts Department of Environmental Quality Engineering (DEQE) for any Federal permit (e.g. Section 404 permit) to conduct an activity which may result in a discharge into waters of the United States. Additional details about the water quality certification process are discussed with the MDPW Environmental Section.

1972 Coastal Zone Management Act — A coastal zone consistency review and statement is required from the Massachusetts Coastal Zone Management (CZM) Office for Federal-aid highway projects located within the designated coastal zone. This review is to ensure consistency with the state coastal zone policies. Additional details about CZM statement process are discussed with the MDPW Environmental Section.

Section 9 of River and Harbor Act of 1899 — A permit is required from the U.S. Coast Guard for highway projects involving bridges or causeways on navigable waters. A water quality certification and a coastal zone consistency statement (if applicable) are needed before the U.S. Coast Guard will issue the Bridge Permit. Additional details about the Coast Guard Bridge Permit process are discussed with the MDPW Environmental Section.

Section 10 of River and Harbor Act 1899 — A permit is required from the U.S. Army Corps of Engineers for highway projects requiring construction in or over navigable waters, the excavation from or dredging or disposal of materials in such waters, or any obstruction or alteration in a navigable water (e.g. stream channelization). Additional details about the Section 10 permit process are discussed with the MDPW Environmental Section.

Table 2-15. Example of state and federal laws and requirements related to highway design (Source: Massachusetts Highway Manual, 1990) continued.....

Section 106 of the 1966 National Historic Preservation Act — A process involving FHWA, the Department, Massachusetts Historical Commission and the Advisory Council on Historic Preservation which must be followed for any Federal-aid highway project affecting bridges, districts, structures, or sites (including archaeological sites) potentially eligible, eligible, or on the National Register of Historic Places. Additional details about the Section 106 process are discussed with the MDPW Environmental Section.

1958 Fish and Wildlife Coordination Act — Consultation is required with the U.S. Fish and Wildlife Service and the agency having jurisdiction over fish and wildlife resources whenever the waters of any stream or other body of water are to be impounded, diverted, or the stream or other body of water otherwise controlled or modified for any purpose on a Federal-aid project. This consultation should be integrated into the Section 404 permit process.

State Laws and Requirements

Massachusetts Wetland Protection Act — This act applies to highway projects whose activities remove, fill, dredge, or alter a resource area defined in the Wetland Regulations or whose activities within the buffer zone of a resource area will alter a resource area. Resource areas are defined as:

(A) Any bank, freshwater wetland, coastal wetland, beach, dune, flat, marsh, or any swamp bordering on the ocean, any estuary, any creek, any river, any stream, any pond, or any lake.

(B) Land under any of the water bodies listed above.

(C) Land subject to tidal action.

(D) Land subject to coastal storm flowage.

(E) Land subject to flooding.

A buffer zone is defined as land within 100 feet horizontally of any resource area listed in (a) above. If the Act applies, then a permit called an Order of Conditions must be issued by the local Conservation Commission. The process of obtaining an Order of Conditions is triggered by an application to do work in a regulated area, followed by a review of the highway project to determine its impact, if any, on the resource area. Work is permitted to go forward only when it can be done in a manner which does not harm the resource involved.

Bridge replacement projects may be statutorially exempt from the Wetland Regulations. Additional details about the very complex State Wetlands Process are discussed with the MDPW Environmental Section.

Chapter 91 — A license is required from DEQE for highway projects that, in coastal or inland areas, involves construction, dredging and filling performed in private and Commonwealth tidelands, as well as great ponds and certain rivers and streams. Additional details about the Chapter 91 process are discussed with the MDPW Environmental Section.

Chapter 152 — A process involving the Department and Massachusetts Historical Commission which must be followed for highway projects affecting bridges, districts, structures or sites (including archaeological sites) on the State Register of Historic places. All properties eligible for or on the National Register of Historic Places are on the State Register. In most cases, the Section 106 process for a Federal-aid highway project will satisfy the requirements of the Chapter 152 process. Additional details about the Chapter 152 process are discussed with the MDPW Environmental Section.

......continued

Table 2-15. Example of state and federal laws and requirements related to highway design (Source: Massachusetts Highway Manual, 1990)

Chapter 3

Application of Geometric Design Principles to Route Design

The fundamental objective in the highway geometric design process is the establishment of the new highway's centerline and cross sections in relation to the terminal points and to the topography through which the highway will pass. The vertical and horizontal alignment of the centerline determines the amount of cut and fill, cross section details, drainage design, construction and user costs, and environmental impacts.

The process of establishing a centerline is described in this chapter by means of examples. The design controls and designations will be assumed in order to illustrate the geometric configurations only. The process of establishing the design controls and designation as a basis for the geometric configuration was discussed in Chapter 2. The principles described are applicable to any classification of highway, from a local access road to a multi-lane freeway. In addition, a brief review of drainage and cost estimates is made, particularly as they pertain to the design projects described in Chapter 5.

PRELIMINARY ROUTE LAYOUT AND GEOMETRIC DESIGN

In selecting a preliminary, technically feasible route, the designer should attempt to envisage the topography in three dimensions. This may be difficult initially, but some practice will assist. The major activities may be divided into: defining design controls; establishing an initial alignment; balancing cut and fill; and, refining the design. These activities are described below. The reader may also wish to consult Chapter 4, where a more formal computational approach is presented as part of a route selection and design project.

Defining Design Controls -- As indicated in Chapter 2, the alignment of a highway is subject to design controls that ensure that it will provide suitable service for the traffic within the topography for which it is designed. As well as the controls noted earlier, it is necessary to specify several other variables that are inputs to establishing a preliminary route. These variables are described as follows:

1. <u>Minimum radius of horizontal curves</u>, based upon the design speed and the permissible superelevation.

2. <u>Minimum length of vertical curves</u>, based upon design speed and difference between intersecting grades.

3. <u>Maximum grade at any point on the highway</u>, determined from consideration of road classification, truck traffic, and terrain.

4. <u>Maximum grade in proximity to existing intersections</u>. The vertical alignment of the proposed route should allow for a minimum to moderate grade approaching the intersection with the existing highway in order to assist in safe stopping on downhills and improved sight distance on uphill approaches. See also the comments in Chapter 2 regarding coordination of horizontal and vertical alignments. Ideally, this grade should be no more than is required for adequate drainage. Because this approach grade may intersect with a vertical curve ascending or descending the hillside, a vertical curve may be required at this location. The length of this curve is determined by consideration of the design speed and intersecting grades. For preliminary design purposes, it is suggested that within a distance of about 30 m from the intersection's stop line, the grade be no more than 2%. This value will be used in the design projects presented later and will allow for any necessary modifications in the detailed design stage.

5. <u>Minimum grade</u> at any point on the highway to ensure adequate drainage. A minimum of 0.5% is suggested for preliminary design purposes.

6. <u>Maximum horizontal approach angle at intersections</u>. It is desirable to design the route to ensure that intersections with existing highways are of suitable alignment and configuration from a safety and capacity point of view. Therefore, the horizontal alignment should feature as nearly as possible a right-angled intersection with the

existing highway (within, say, $90^o \pm 15^o$). It is suggested that for preliminary design purposes, the proposed approach to the intersection be a tangent section for a distance of at least 30 m to aid drivers' visibility at the approach.

7. <u>Maximum depth of excavation and height of fill</u>. For the reasons specified in Chapter 2, a maximum depth of cut and height of embankment must be specified in order for the designer to establish an initial vertical alignment.

Establishing an Initial Alignment -- Development of the alignment is a trial and error process involving defining a trial alignment, then checking to see if it complies with the horizontal and vertical controls, then modifying it in successive iterations until all the controls are complied with. One approach to this process is illustrated by the problem example shown in Figure 3-1. In addition to these steps, the following points may help to guide the process.

<u>Horizontal Alignment</u>. A first step is usually to determine if the shortest route possible will comply with the controls, because this is likely to be the least-cost solution. Examination of how this first trial route complies with the controls will suggest how the route may be modified for the next trial. The highway should be constructed as close to the existing ground (or slightly above it to assist adequate drainage) as possible, provided that the design controls are complied with. Thus, any horizontal centerline should be checked, first of all, for its grade. This may be done approximately by measuring the length of a given segment of highway and counting the contour lines that are crossed. The vertical distance covered, divided by the horizontal length, indicates the approximate grade. If this grade is significantly more than the specified amount, the alignment must be readjusted.

Where the rounded topography of a mountain or a hill must be negotiated in a transverse fashion, the curve of a highway should preferably conform approximately to the surface of the hill itself, or excessive cuts or fills are likely to result. At this point, the designer must sketch a curve that approximately conforms to the topography, by using compasses or templates. This curve must then be checked for conformity with the maximum allowable radius and also for the grade the highway negotiates throughout the curve. The latter must also conform with the maximum allowable grade requirement and be adjusted if necessary.

FIGURE 3-1
OUTLINE OF STEPS IN DETERMINING A POTENTIAL HORIZONTAL ALIGNMENT

PROBLEM: Connect points A and B below with a highway having a maximum grade of 6%, minimum grade (for drainage) of 0.5%, a minimum horizontal radius of 240 m, a minimum vertical curve length of 120 m, and maximum cut and fill depths of 10 m.

STEP 1. Examine the contours along the shortest possible route (hypothetical) and estimate the steepest ground slope along this route. The steepest slope is between points X and Y (from examination of contour lines) - a vertical rise of nearly 38 m in 240 m horizontally. This represents a slope of over 15%. This is more than twice the allowable maximum grade of 6% and would result in excessive depths of cut and heights of fill. Therefore a less steep route should be examined, as shown in Step 2, below.

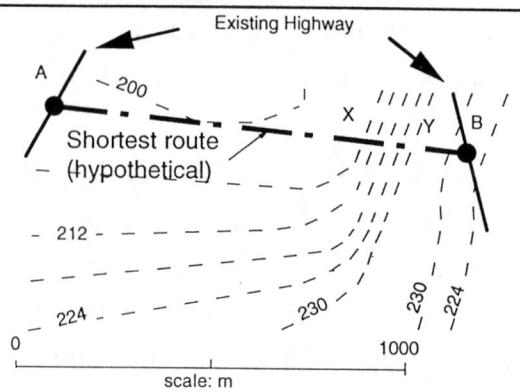

STEP 2: Sketch a new trial alignment with a reduced ground slope. The one shown here is sketched freehand and includes a curve, the radius of which is approximately the same or larger than the 240 m minimum specified. Notice the grade between points X and Y of Step 1 will be less steep because the horizontal alignement of the highway crosses the contour lines at a greater angle than in Step 1.

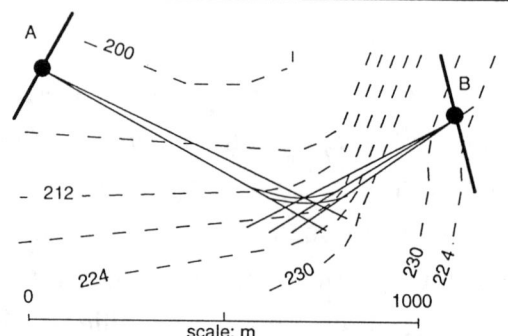

STEP 3: Convert the sketch of Step 2 into a dimensioned tangent and circular curve alignment by scaling and the use of compasses or a template for constructing and measuring the curve. It is also usually desirable to ensure that the intersection of the proposed highway with the existing road is within about 15 degrees of a right-angle in order to assist driver sight distance requirements.

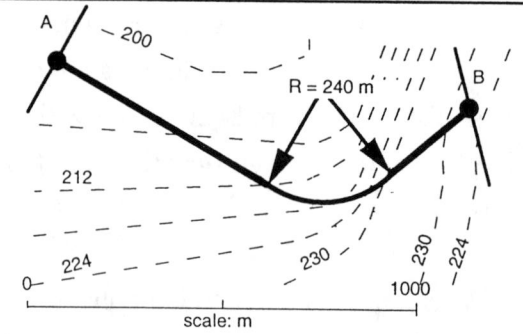

STEP 4: Construct a profile of the ground levels as a basis for defining a profile of the highway pavement which does not exceed the 6% slope and the 10 m cut and fill requirements. This is usually done to balance the cut and fill amounts also, and an acceptable (though not necessarily unique) solution is shown here, for the alignment of Step 3 above. It is also usually desirable to ensure that the grade of the proposed road where it intersects the existing road is a maximum of about 2% to assist stopping and sight distance requirements. Note also that the minimum grade requirement of 0.5% has been complied with. The diagram indicates the combination of sag vertical curve, (svc) straight grades, and crest vertical curve (cvc) that together comprise the total length of the potential route. The total length of the route is 1261 m.

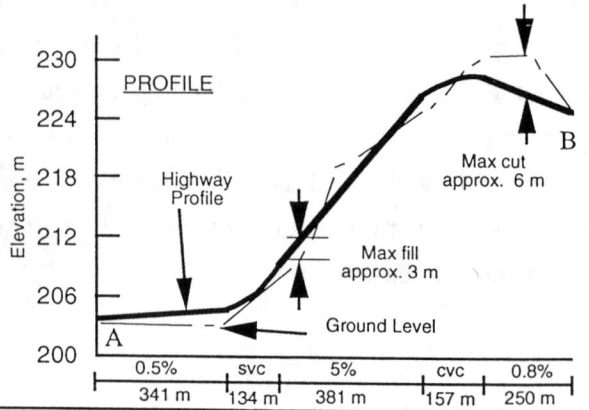

Existing Ground Profile and Vertical Alignment. Once a preliminary route has been defined using the above steps, the next step is to examine the profile. This means first drawing a longitudinal section of the existing ground level along the horizontal alignment. To do this quickly but approximately, a strip of paper may be laid along the centerline of the proposed highway. The contour line elevations along the alignment are then marked off along the edge of the paper and the results are transferred to the profile sheet to provide a profile of the ground level along the proposed route. This is an approximate method but it will save considerable time and can be refined later. A more accurate method is to use dividers or an engineer's scale to measure the horizontal distance and transfer these measurements and the elevations to the profile. Care should be taken to ensure that the horizontal distances between contour lines are measured accurately -- especially along curves.

Balancing Cut and Fill. If the vertical profile mentioned above meets all of the design controls it should now be checked to see if the cuts and the fills approximately balance - to ensure that excessive amounts of soil need not be imported to or removed from the site. Any adjustment of the grade may be done initially by visual, trial and error means using a straight edge and a circular or French curve together with approximate measurements at peaks and valleys to ensure that the maximum cut and fill dimensions along the centerline are not exceeded and that all grades and curve lengths are complied with, based upon the appropriate stopping sight distance. Also, it is desirable where possible to minimize the amount of cut and fill, as well as the amount of uphill haul of fill material. These requirements may be addressed by modifying the initial profile. Note that it may not be possible to balance the cuts and fills while still conforming with the other design controls. Nevertheless, the horizontal and vertical alignments should be adjusted to obtain the best possible balance.

Note: The vertical scale of the profile is usually exaggerated on drawings to provide a better visual image and permit scaling of cuts and fills. Also, it is often useful to draw to one side of the profile the maximum grade and cut and fill dimensions as an aid to the sketching process. The suggested requirements for the horizontal alignment and vertical grades adjacent to intersections are shown graphically in the Appendices.

Refinements to Selected Route -- Once an initial, technically feasible route has been defined and examined, the alignment may be adjusted to ensure that the relevant K values have been complied with, address coordination of horizontal and vertical curves, and explore other routes that involve, for example, less depth of cut or height of fill or reduce the proximity to sensitive features such as wetlands.

The alignment selection process can now be followed for a number of alternatives. There are several reasons for doing this. For example, the shortest highway that is feasible in a technical sense may not be the least expensive. Detailed economic analysis will be needed to determine these relative costs. Also, in practice, the provision and estimation of several alternatives will provide information for decisionmakers who may favor certain alignments over that considered preferable by purely engineering evaluation. If possible, at least three alternatives should be initially defined, all of which are technically feasible and conform to the specified design controls.

EXAMPLE OF DEVELOPING AND CHECKING ALTERNATIVE ALIGNMENTS

In applying the procedures outlined above, we now examine the main features of developing alternative alignments through a particular topographic area. The details for this particular example are described as follows:

1. We wish to make a preliminary analysis of a highway route connecting points A and B shown in Figure 3-2.

2. A design speed of 80 km/h and a maximum allowable superelevation of 10% have been specified. Thus, the minimum allowable horizontal curve radius is 210 m, as indicated in Table 2-8.

3. The maximum allowable grade is 10% (except as specified for the areas adjacent to intersections), based upon the anticipated vehicle types.

4. The horizontal intersection angle and the maximum grade of the proposed highway at the intersection are as described earlier (horizontal intersection angle to be 90±15 degrees within 30 m of an intersection, maximum grade to be ±2% for a distance of 30 m from the existing highway, minimum grade at all locations ±0.5%).

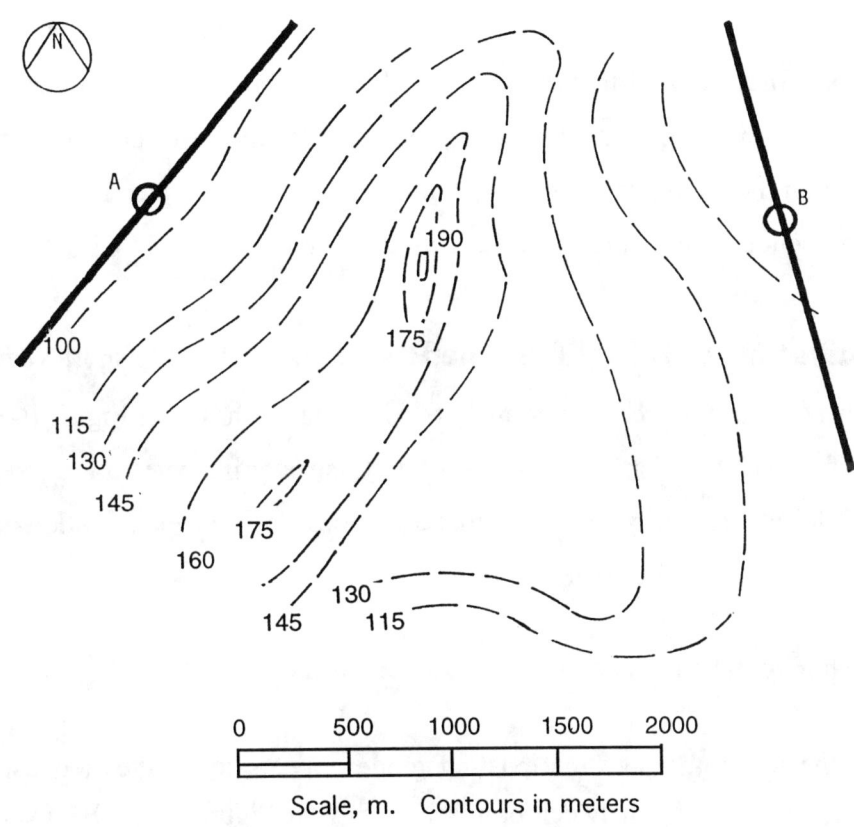

FIGURE 3-2

**LOCATION OF END POINTS
OF PROPOSED HIGHWAY**

5. The minimum allowable vertical curve lengths are obtained from Figures 2-8 and 2-9. The "worst case" crest curve length will be where a 6% positive grade intersects a 6% negative grade, giving a length of 590 m (A = 12%, V = 80 km/h, Figure 2-8). Correspondingly, from Figure 2-9, the "worst case" minimum sag curve length is 385 m. At this point we are not considering the K value for drainage purposes, which would normally be done before the alignment is finalized.

6. No environmental/cultural features are described in this example in order to focus on the technical design requirements only. The presence of these features would impose constraints on the alignment, similar to those described in Chapter 1 and in the section on environmental impacts presented later in this chapter.

Investigating an Initial Possible Route -- A reasonably short route for which we might wish to make a preliminary analysis is Route 1, shown in Figure 3-3A. Even though a brief visual check on the contours along this route indicates a ground slope well in excess of the 6% specified, we will make a more detailed examination in order to illustrate the problems. The steps are as follows:

1. Draw the horizontal alignment to scale along the route.

2. Examine the location where the existing grade appears to be the steepest along the route. This is obviously between points X and Y in Figure 3-3A. By examining the horizontal scale and the vertical contour interval, we can see that between points X and Y the existing grade is approximately 30 m/300 m = 10%. Clearly, this exceeds our maximum allowable grade of 6% for some considerable horizontal distance. However, we check by sketching a profile to scale in order to examine the depth of cut and fill and, therefore, the practicality of the route, in more detail (Step 3 below).

3. By drawing a ground profile along Route 1, together with a profile of the centerline of the highway with a maximum of 6% grade, as shown in Figure 3-3B, it is obvious that this route will result in an unacceptable cut (over 15 m deep) and nowhere to use the cut material as fill. Even with a slight adjustment of the alignment, it is apparent that no satisfactory improvement in these cuts and fills will be possible for this route.

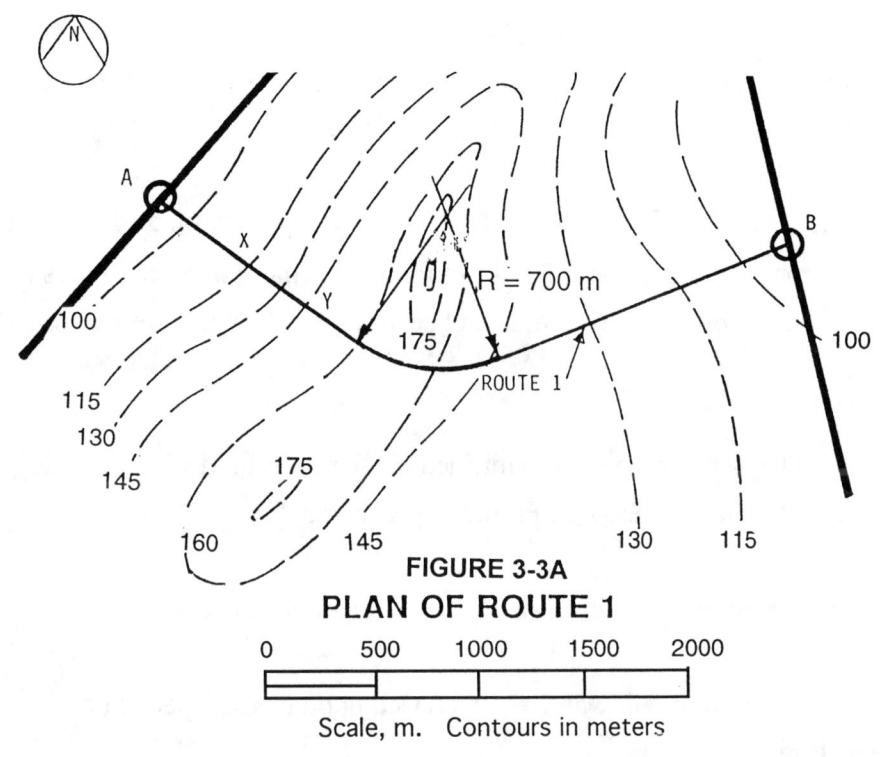

FIGURE 3-3A
PLAN OF ROUTE 1

| 0 | 500 | 1000 | 1500 | 2000 |

Scale, m.   Contours in meters

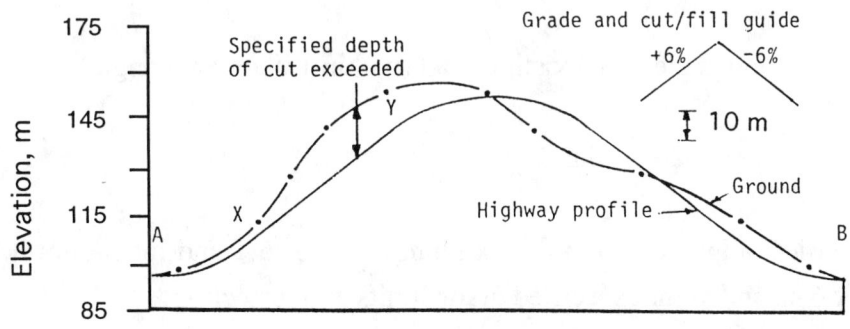

FIGURE 3-3B

PROFILE ALONG ROUTE 1

4. Summarize conclusions as follows:

   a) The inability of Route 1 to provide a gradient within the required 6%, while simultaneously satisfying the cut and fill requirements, makes Route 1 and any adjacent or somewhat similar route unacceptable alternatives.

   b) Knowing the implications of Route 1 in terms of cut, fill, and allowable grade, it is therefore necessary to explore several other alternative routes to attempt to establish a technically acceptable alignment.

Investigation of Routes 2 and 3 -- After examination and preliminary sketching, it is apparent that Routes 2 and 3 might offer more gradual grades and be worth investigating. These routes are shown sketched in Figure 3-4 and are examined in greater detail below.

Route 2. Using the procedures outlined earlier as a guide for checking the route's technical feasibility, and as shown in Figures 3-5A and 3-5B,

1. Convert the sketch of Route 2 into a series of tangents and curves.

2. Check for the minimum allowable radius based upon design speed and superelevation.

3. Check for intersection angle with existing road (within 15° of right angle).

4. Construct the existing grade profile.

5. Establish a vertical alignment with a maximum grade of 6% and maximum height of cut and fill of 6 m, and within specified grade limits at intersections.

Judging by the design controls established earlier, it can be seen from Figures 3-5A and 3-5B that Route 2 is a technically feasible alternative. Also, the cuts and fills appear to balance fairly well. The horizontal and vertical alignment could, of course, be adjusted slightly and each engineer will arrive at a slightly different geometric design, at least from these preliminary efforts.

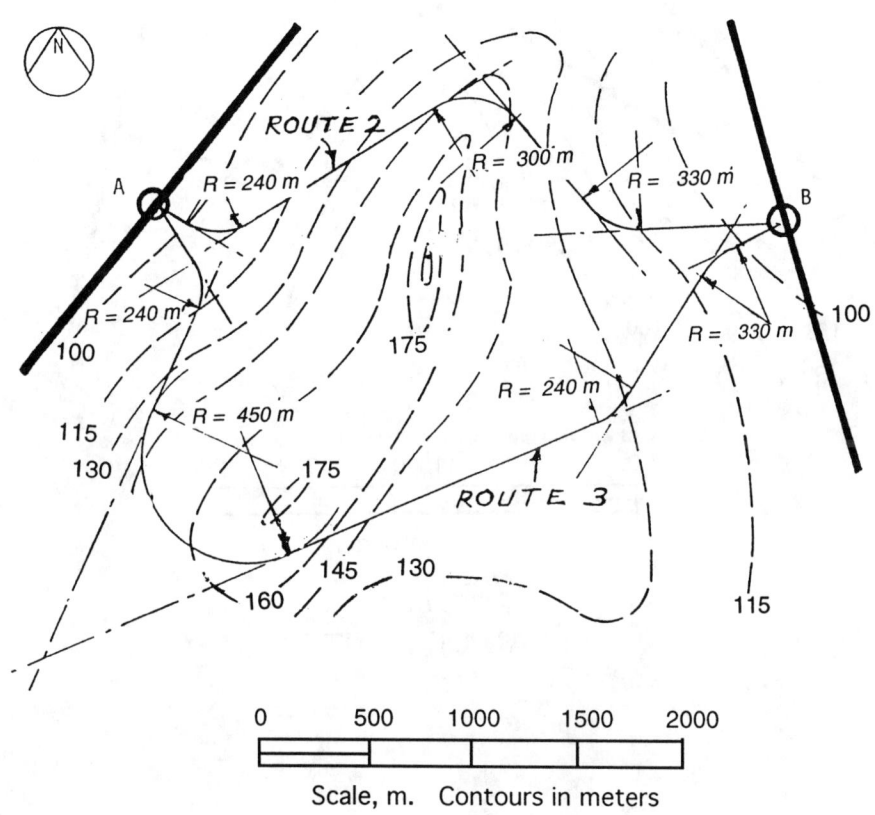

Note: Tangents and curves are sketched to show
approximate route alinements only.

FIGURE 3-4

INITIAL DEVELOPMENT SKETCHES OF ROUTES 2 AND 3 RESULTING FROM EXAMINATION OF ROUTE 1

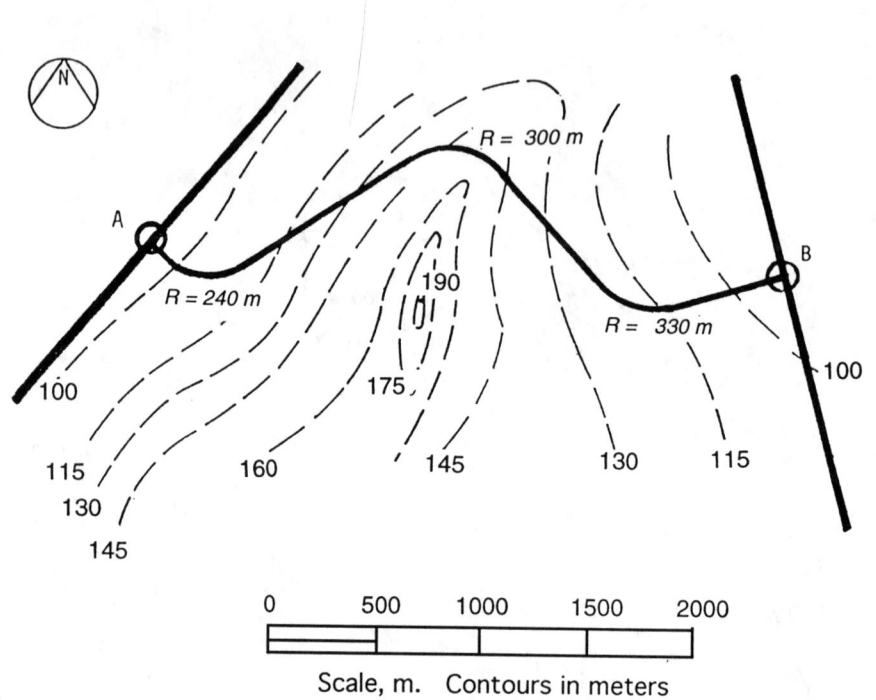

R = 300 m

R = 240 m

R = 330 m

A

B

N

190

175

160

145

130

115

100

130

145

115

130

145

160

175

190

130

115

100

0   500   1000   1500   2000

Scale, m.   Contours in meters

FIGURE 3 -5A

PLAN OF ROUTE 2

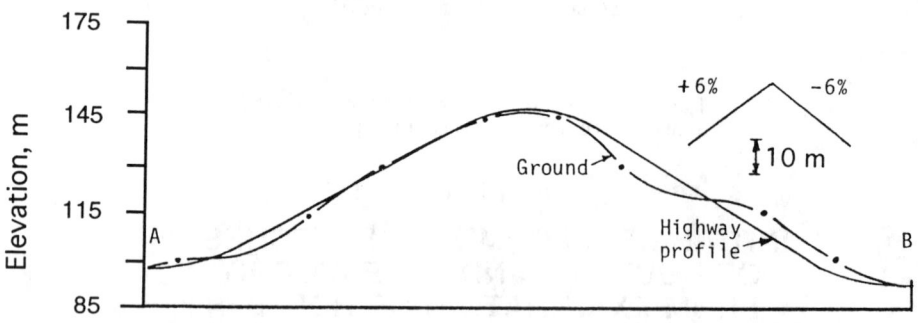

175

145

115

85

Elevation, m

A

B

Ground

Highway
profile

+ 6%   − 6%

⊥10 m

FIGURE 3-5B

PROFILE ALONG ROUTE 2

<u>Route 3</u>. Using the same approach as that used for Route 2, an alignment for Route 3 is developed from the initial sketch. The procedure for developing Route 3 is shown in Figures 3-6A and 3-6B, indicating that Route 3 is a technically feasible route also.

Screening and Selection of Routes for Preliminary Design -- Both Routes 2 and 3 appear to be technically feasible, based upon the allowable grade, cut and fill depths, and horizontal and vertical alignments, while Route 1 clearly is inadequate. It is often useful to screen the proposed routes at this point in order to summarize in concise form the reasons why one or another route should be considered further. Table 3-1 lists a number of major criteria and comments on how each route meets each criterion. The conclusion, as indicated above, is that both Routes 2 and 3 are technically feasible and that a preliminary design and economic analysis should be conducted as a basis for determining the prefered alternative. The three routes investigated are depicted in Figure 3-7.

Highway Centerline Traverse -- At the current stage of the design (i.e., development of a preliminary alignment), the intersecting angles and centerline dimensions may be scaled, but the traverse should "close" at least approximately so that the data given to a field survey party will be adequate for performing a more detailed ground survey. The centerline dimensions and intersecting angles, together, provide a check on the traverse angles and distances to ensure that "closure" occurs, (i.e., that the beginning and end points coincide within a reasonable degree of accuracy). This process is described in basic texts on surveying and is not discussed further here.

Important Note: Particularly when maximum depths of cut and height of fill are specified, it may not be possible to obtain an alignment which conforms to the design designation and controls. In these cases, the designer must decide if bridges or tunnels will be permitted or if controls on grade, design speed, or other determinants of the alignment can be relaxed.

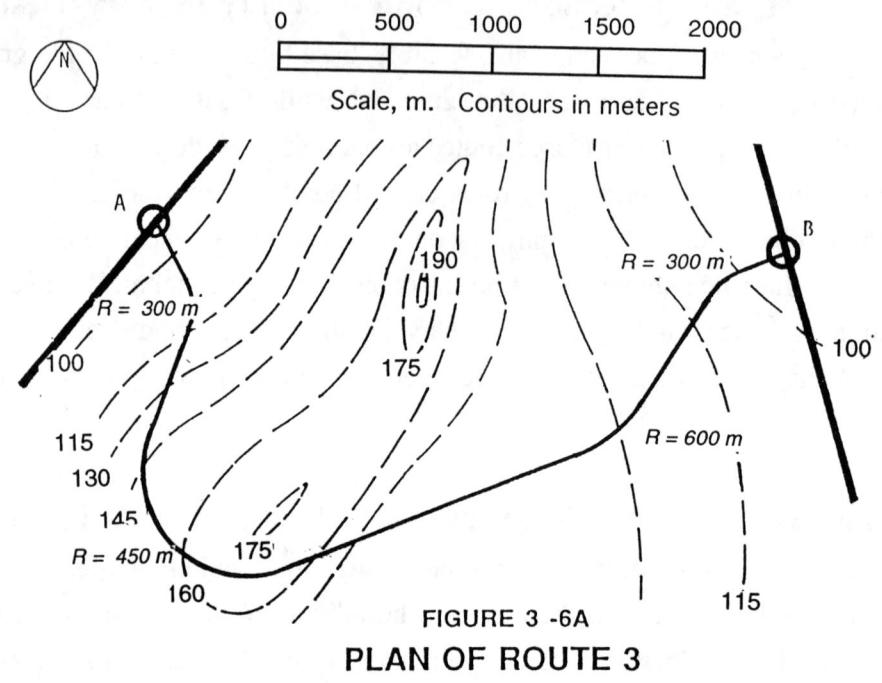

FIGURE 3 -6A

PLAN OF ROUTE 3

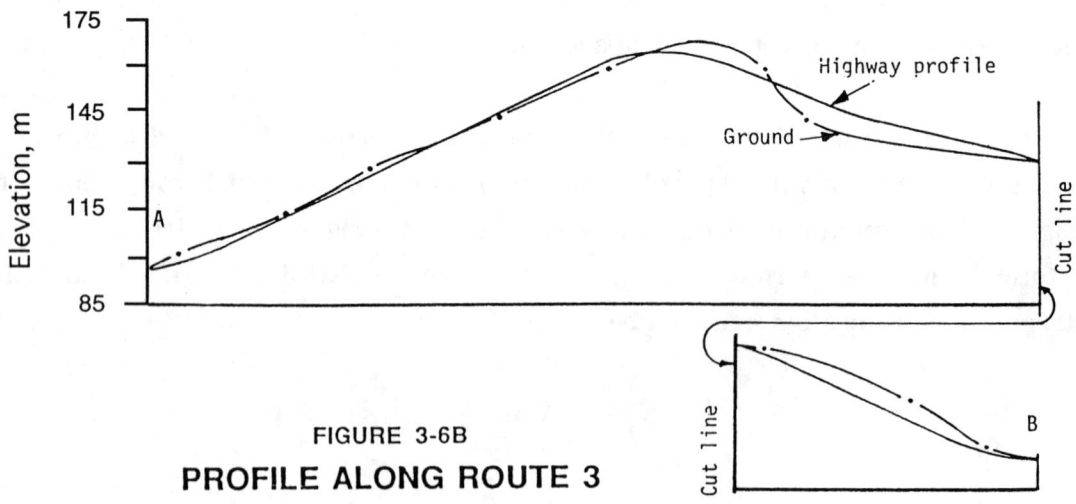

FIGURE 3-6B

PROFILE ALONG ROUTE 3

CRITERIA	SCREENING EVALUATION		
	ROUTE 1	ROUTE 2	ROUTE 3
Length of route (approximate)	3030 m	3540 m	4180 m
Conformance with design controls	Not possible with specified grade control	Acceptable	Acceptable
Cut and fill balance	Excessive cut required to comply with design controls	Acceptable	Acceptable
Need for bridges or other special structures	None	None	None
Environmental impacts	Excessive cuts and associated slopes	No essential difference between Routes 2 and 3	
Potential high cost items	Excessive cuts and fills	None evident	None evident
Minimize total cut and fill and minimize uphill haul	Some uphill haul is likely with each alternative		

Conclusion:
Route 1 is unacceptable due to the need for excessive excavation required to attain the specified grade control. _Routes 2 and 3 appear technically feasible and should be further investigated by means of an initial economic analysis before a detailed design is undertaken_

TABLE 3-1

SCREENING EVALUATION OF
ALTERNATIVES

ROUTE 2
ROUTE 1
ROUTE 3

A

B

FIGURE 3-7

OBLIQUE VIEW OF
ALTERNATIVE ROUTES 1, 2 AND 3

NON-STANDARD SITUATIONS

Particularly in mountainous terrain, it may often be the case that an acceptable alignment that conforms to the specified controls is difficult to attain without extreme measures such as deep cuts, use of bridges or even tunnels, particularly where the highway must traverse a number of valleys. Usually the solution entails either provision of horizontal curves with radii less than the allowable, and associated speed restrictions, or the provision of bridges. In cases where these design alternatives exist, a more detailed analysis must be carried out, yet the principles described earlier apply. An example of how a bridge may provide a better solution than a horizontal curve of substandard radius is shown in Figure 3-8. Again, the final decision will rest upon construction, maintenance, and user cost estimates and comparisons.

DRAINAGE PROVISIONS

An initial drainage design indicating the main locations of catchment areas, ditches, culverts, and bridges is an important part of the preliminary highway design because the alignment may have to be changed if the road cannot be adequately drained, or if it adversely affects existing drainage patterns.

The identification of runoff areas likely to affect the highway geometric design (particularly the horizontal and vertical alignments) is of crucial importance for a satisfactory design. The highway, as well as being affected by the characteristics of the watershed such as slope and ground conditions, will itself affect the flow of surface and, perhaps, subsurface drainage in its vicinity. The provision of adequate drainage ditches, culverts, and bridges is therefore of vital importance. See the bibliography in Chapter 1 for a selection of drainage-related guidelines

One way of conducting a preliminary drainage design is to define the characteristics of the major precipitation catchment areas; estimate quantities of runoff; locate ditches, culverts, and bridges; check several "worst case" ditch, culvert, and bridge dimensions; and ensure that adjacent drainage patterns of the surrounding topography are not adversely affected by changes in flow patterns. This process may be complex, depending on the location, topography, ground conditions, and environmental factors. The reader should consult the appropriate texts and manuals and, wherever possible, obtain first-hand knowledge of local practices and conditions. A preliminary drainage design may be made, however, to the extent necessary to define the basic configuration, dimensions, and construction costs and to indicate where a field survey crew should examine various features in greater detail. An example of this approach is included in the project described in Chapter 4, and examples of typical drainage facilities related to terrain and highway characteristics are shown in Figure 3-9.

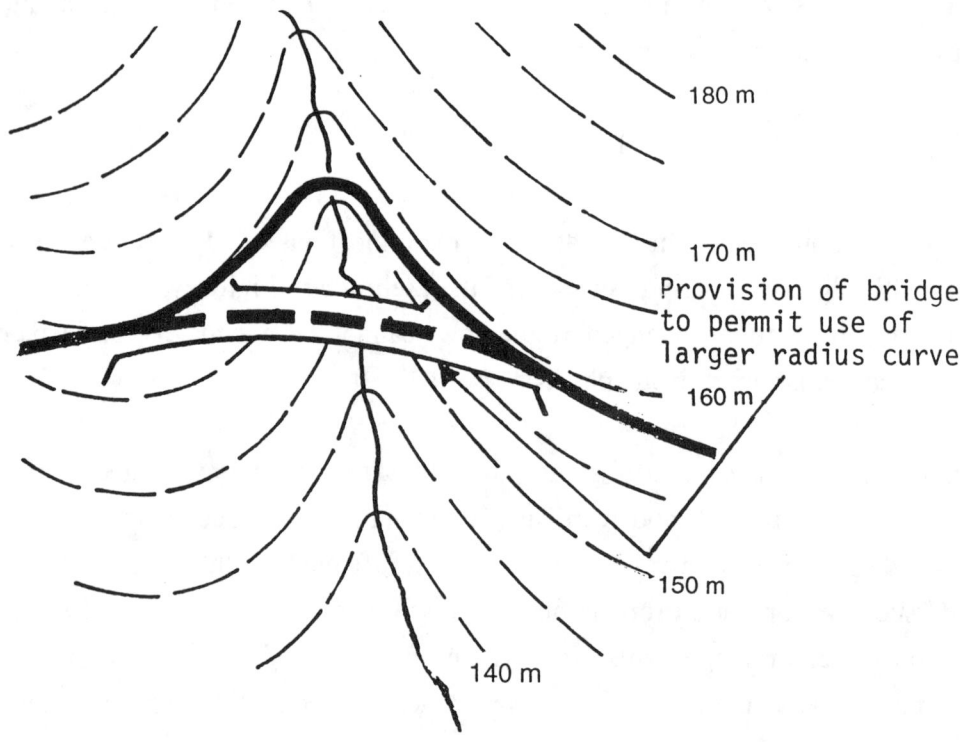

180 m

170 m

Provision of bridge to permit use of larger radius curve

160 m

150 m

140 m

FIGURE 3-8

EXAMPLE OF WHERE A BRIDGE MAY HELP TO IMPROVE GEOMETRICS

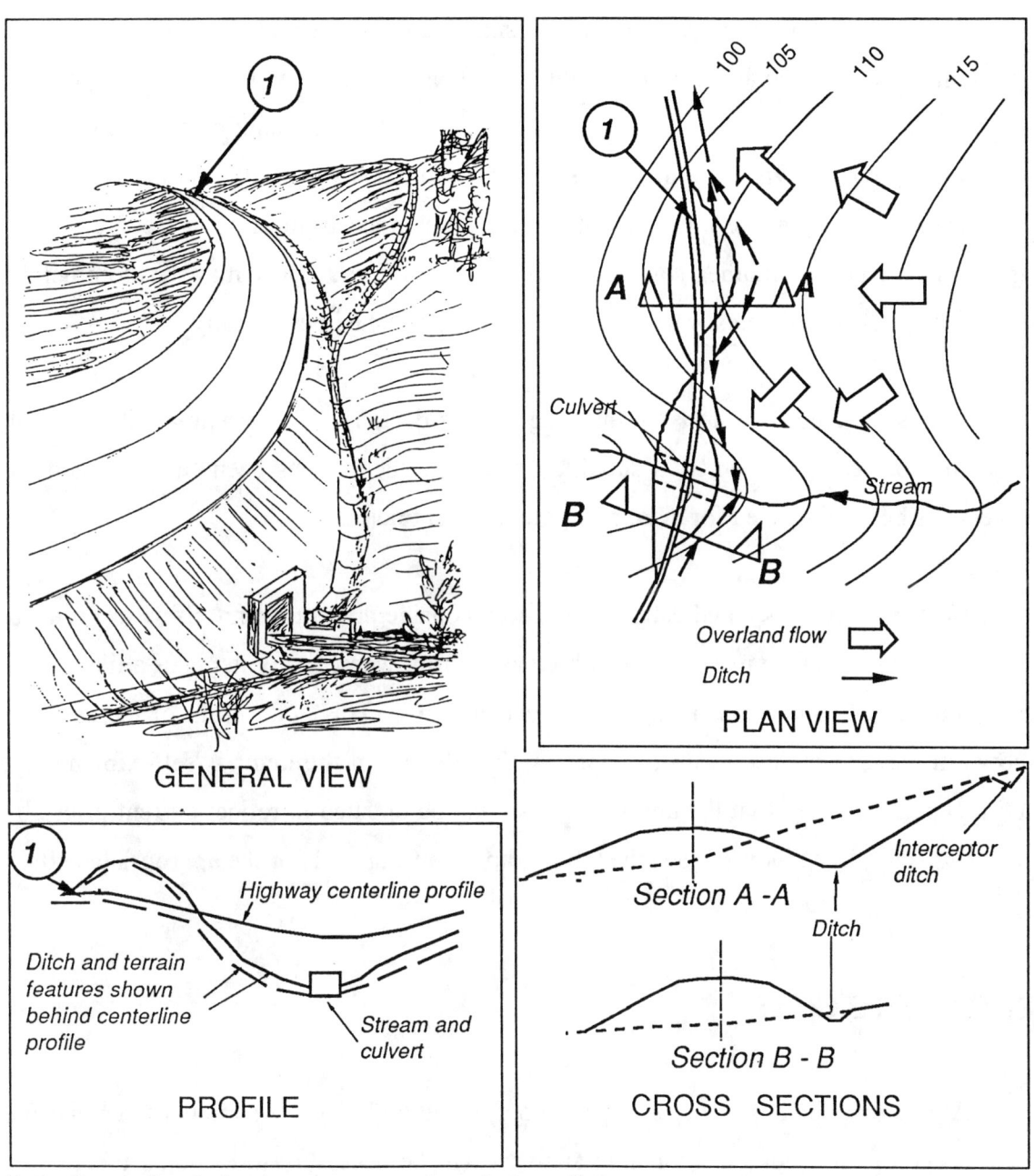

GENERAL VIEW

PLAN VIEW

Overland flow ⟹

Ditch →

Culvert

Stream

A △ △ *A*

B △ △

B

100 105 110 115

PROFILE

Highway centerline profile

Ditch and terrain
features shown
behind centerline
profile

Stream and
culvert

CROSS SECTIONS

Section A -A

Section B - B

Interceptor
ditch

Ditch

Note: Guardrail and other features are not shown in these diagrams.
Diagrams are not to scale.

FIGURE 3 -9

**EXAMPLES OF DRAINAGE FEATURES -
DIAGRAMMATIC**

CONSTRUCTION COST ESTIMATE

For most preliminary engineering designs, and as indicated in Chapter 1, an approximate cost estimate will be required. This estimate will be based upon the quantities of materials, labor, equipment, overhead, and profit required to construct the proposed highway. While the quantities may be estimated in the appropriate units from the preliminary design, the unit cost rates may be obtained from previous records of the agency concerned. Another source of information of unit rates may be commercial sources of construction rates which are updated and published periodically.

It must be emphasized that at this stage of the design, the cost estimate is not final, nor is it a bid estimate, which is submitted when the design has been finalized and the project has been advertised for bids from potential contractors.

Because a fully detailed estimate involves considerable time and effort to produce, the preliminary estimate may take an abbreviated form, where many items of a like nature are "compressed" into a much smaller number of items. This is the form of the estimate shown in the project described in Chapter 4. The listing of the items is self-explanatory, and it should be noted that the unit cost rates must be updated to reflect current prices by means of appropriate indices. See the bibliography in Chapter 1 for the appropriate texts.

ECONOMIC COST

Although estimates of the economic cost of the highway are somewhat beyond the main scope of this book, an economic analysis will be necessary to estimate the cost of vehicle operations and of vehicle occupants' travel time, in addition to the capital and maintenance costs of the highway. As indicated earlier, a project may be technically feasible but not economically justifiable, and this book focuses on the former concern.

The usual method of conducting the economic cost analysis is by means of the benefit-cost and present value techniques described in the AASHTO "Red Book" (see bibliography, Chapter 1). Using this method can be time consuming. However, a simplified means of comparing the relative costs of various alignments can be made in order to illustrate the principles involved, and to assist in a basic economic comparison between projects. The main features of the method were described in Chapter 2, and a simplified economic cost estimate to permit comparisons between projects is described as a part of the project in Chapter 4.

ENVIRONMENTAL IMPACT ANALYSIS

In addition to the wider issues associated with the presence of the highway within the overall transportation system, its detailed locational features must be sensitive to local and adjacent features likely to be impacted. This may require modifications to the alignment to avoid wetlands, habitats of endangered species, and other natural or man-made features of social or cultural significance.

Therefore, in selecting and formalizing the alignment, the curvature and grades outlined in earlier examples in this chapter may have required additional constraints if environmentally sensitive areas had been present nearby. Examples of the relationship of environmental features adjacent to a proposed route are shown in Figures 3-10 and 3-11. The examples also include the written description for the specific items mentioned.

Although a detailed environmental report as indicated in Chapter 2 requires considerable detailed information, the design shown in Chapter 4 mentions the key points and mitigation features only. This is intended to provide an initial alert to the environmental planners to the more salient items of which the highway engineer is aware.

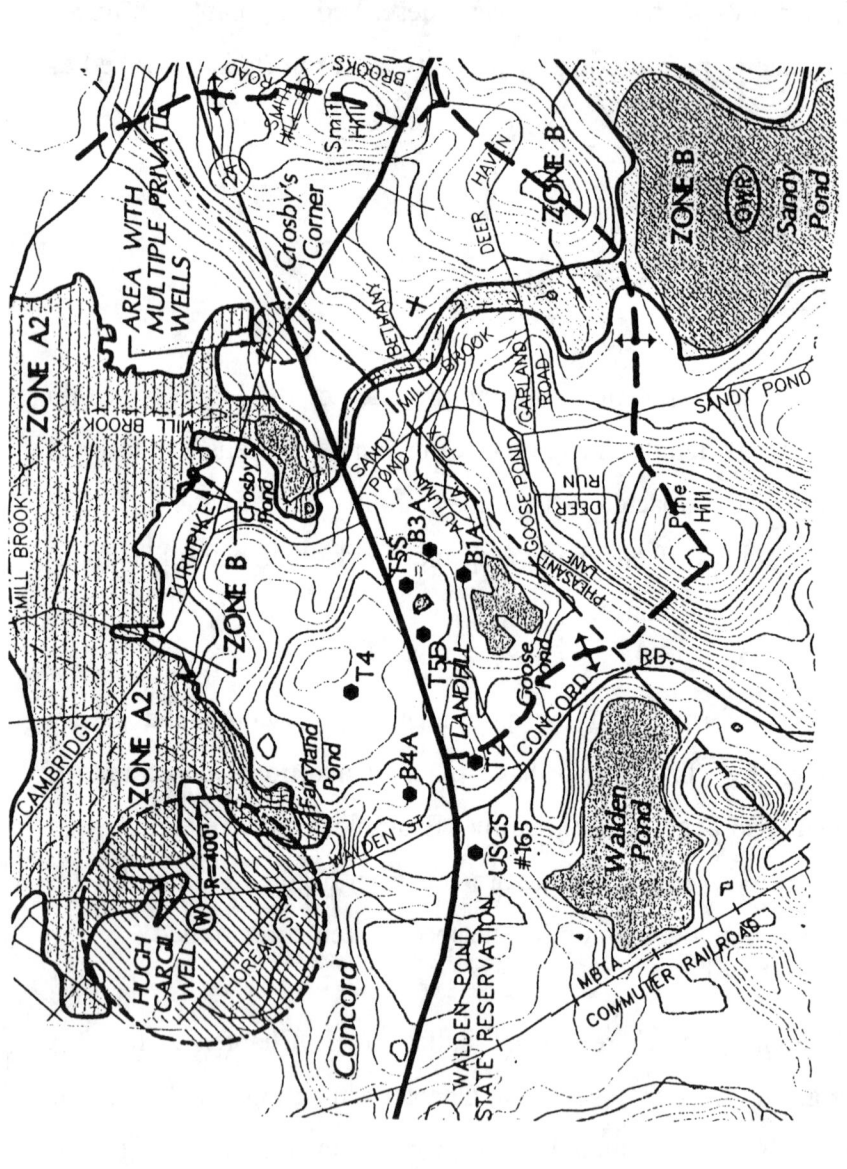

FIGURE 3-10

EXAMPLE OF WETLAND LOCATIONS AND CATEGORIES

(SOURCE: DRAFT ENVIRONMENTAL IMPACT REPORT/ENVIRONMENTAL ASSESSMENT AND
SECTION 4(f) EVALUATION, ROUTE 2 CROSBY'S CORNER, MASSACHUSETTS HIGHWAY
DEPARTMENT, OCTOBER, 1998)

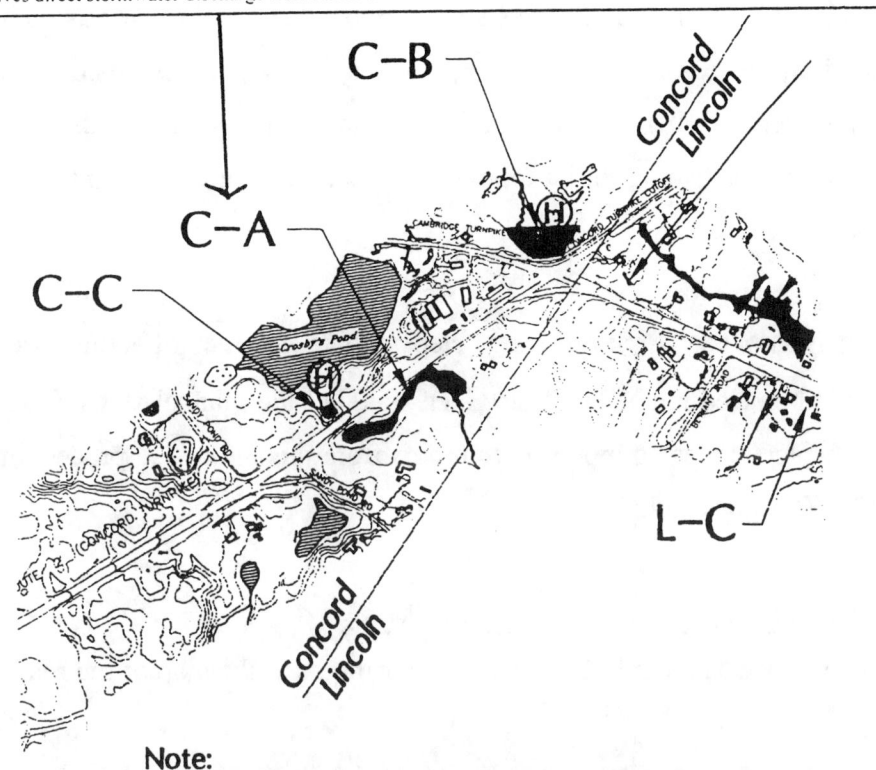

Wetland C-A: Wetland C-A is a bordering vegetated wetland adjacent to Mill Brook, a tributary stream to Crosby's Pond south of the eastbound lane of Route 2, across from the MassHighway garage. The stream enters the project area from the south and parallels the existing Route 2 for approximately 800 meters (0.5 mile) before passing through a culvert under Route 2 to the pond. The canopy of this wetland is dominated by red maples. The understory vegetation in Wetland C-A consists of highbush blueberry, spicebush, arrowwood, and swamp azalea. Ground cover is predominately skunk cabbage, with areas of poison ivy. The stream in this wetland receives direct stormwater discharge from Route 2.

Note:
C Designates wetland located in Town of Concord
L Designates wetland located in Town of Lincoln
(H) Mystic Valley amphipod (Crangonyx aberrans) habitat

FIGURE 3-11

EXAMPLE OF WETLAND DETAILED DESCRIPTION
(SOURCE: DRAFT ENVIRONMENTAL IMPACT REPORT/ENVIRONMENTAL ASSESSMENT AND
SECTION 4(f) EVALUATION, ROUTE 2 CROSBY'S CORNER, MASSACHUSETTS HIGHWAY
DEPARTMENT, OCTOBER, 1998)

AUTOMATED GEOMETRIC DESIGN

Current Status of Automated Geometric Design -- Determination of an appropriate route in accordance with the necessary design specifications and controls have been the subject of mostly proprietary automated methods of design since the 1980's. Most methods take advantage of, and may be embedded within, various computer-aided design and drafting (CADD) design approaches. Also, the inclusion of geographic information systems (GIS) information storage and retrieval methods often plays an extensive role in establishing initial topographic features along the highway corridor and its environs.

Design Process -- Several key steps typify conduct of a highway geometric design using automated methods. These steps parallel the manual steps described in this and earlier chapters. Selected inputs and outputs from the automated process are shown in Figure 3-12. The steps include:

1. Construction of Digital Terrain Model (DTM)

- Input of survey data records and updates – for inputs to the map of the road corridor

- Completion of maps at various scales and details, including relevant surface and subsurface information if required.

- Construction of digital terrain model (DTM) showing all vertically and horizontally defined features, including contours. The DTM is linked to a triangulated irregular network (TIN). The DTM model can be portrayed in three dimensions.

2. Plotting Alternative Routes

- Input of standard features such as cross sections templates, fill slopes, and design constraints such as maximum grades, depths of fill, minimum curvature, and related features.

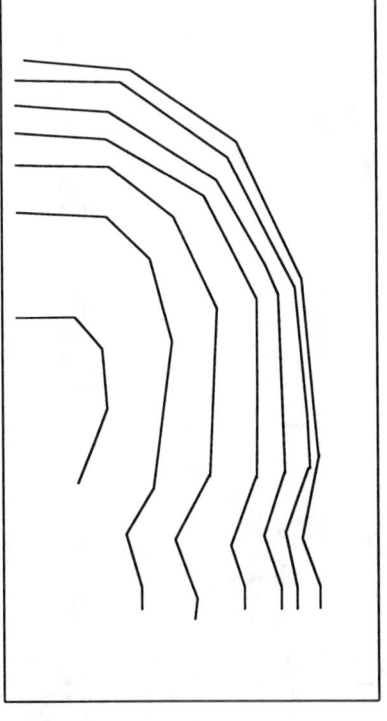

2. ...contours are defined

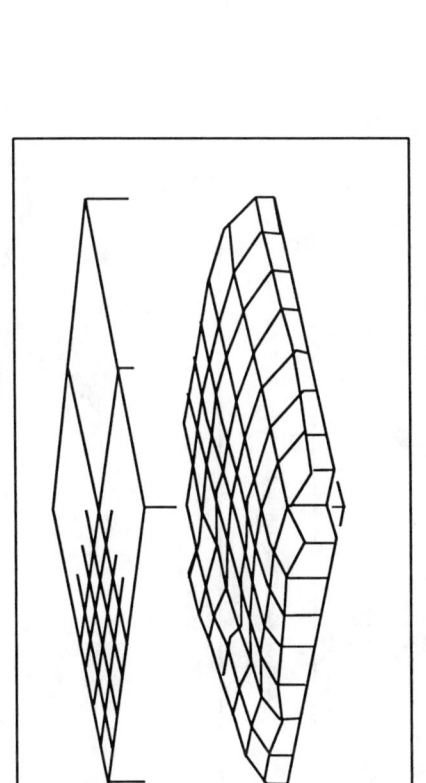

4. ...an initial route is selected based upon results of step 3, and on geometric design controls and environmental factors

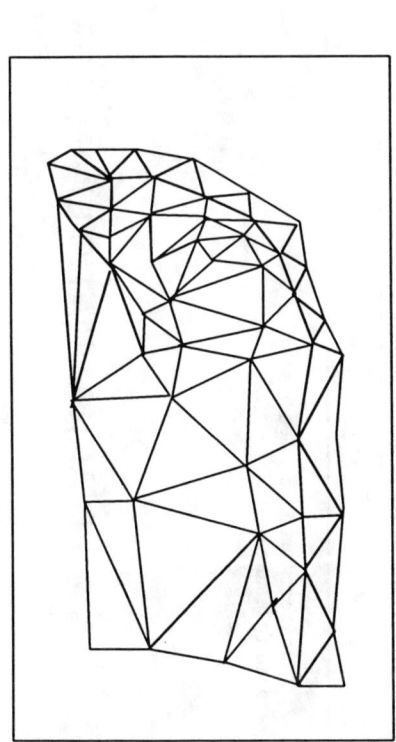

1. A triangulated irregular network (TIN) is prepared to establish corridor-wide control points

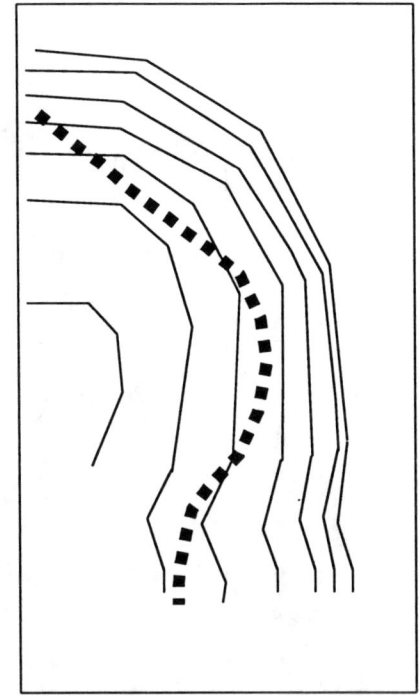

3. ....a 3-dimensional grid is established

FIGURE 3-12

DIAGRAMMATIC EXAMPLES OF ITERARATIVE STEPS IN AUTOMATED ROUTE DESIGN

Continued.....

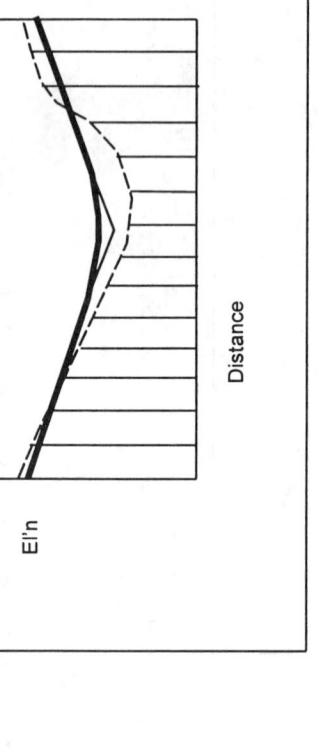

El'n

Distance

6. ...an acceptable profile is determined

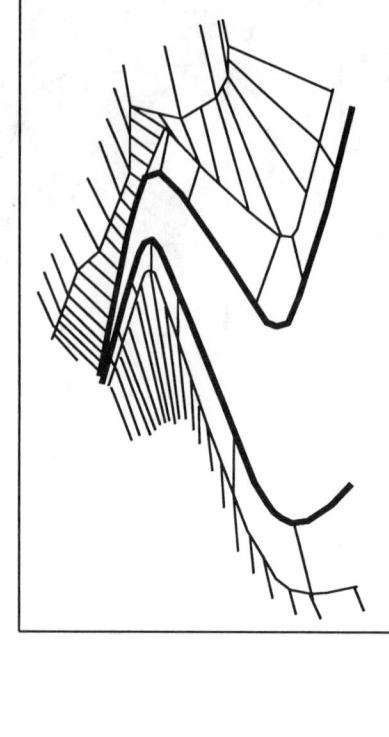

8. ....an oblique digitised route representation is
created to assist in investigating alternative routes
based upon cost and environmental factors

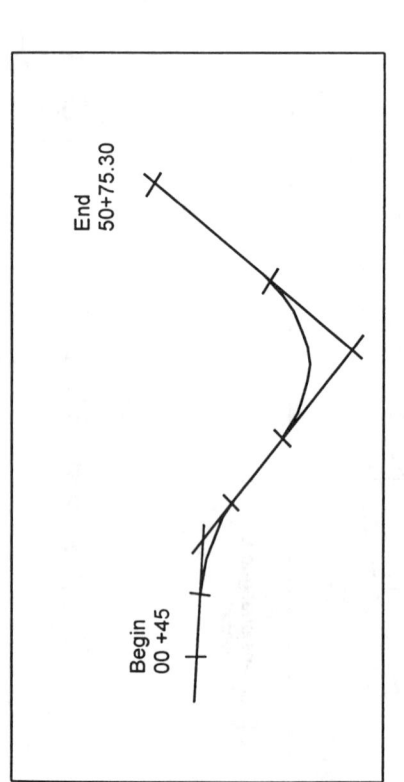

End
50+75.30

Begin
00 +45

5. ...a dimensioned horizontal alignment is established

7. ...cross sections are detailed based upon the profile
and terrain features

...continued, FIGURE 3-12

DIAGRAMMATIC EXAMPLES OF ITERATIVE STEPS IN AUTOMATED ROUTE DESIGN

- Automated estimation of alternative horizontal alignment, profile, earthwork quantities, and major geometric features, to enable interactive editing and re-design until acceptable alternatives have been identified.

- Generation of 3-D, oblique, and "drive-thro" portrayals of selected views of the route to assist better understanding of the route's impacts and essential construction mitigation measures. These graphical portrayals can assist considerably in understanding the implications of the project – particularly by lay people and decisionmakers.

- Generation of design drawings for approval, bids, and construction.

At present, use of automated design features continues to be relatively expensive, both in terms of the equipment used, its connection with GIS systems, and the needed expertise to employ it efficiently. Consequently, many states do not have sufficient projects to justify these costs and much of the work is carried out by consulting firms. As costs decrease, greater use of these methods is likely. Yet it must be remembered that although their use can considerably speed the design process, the basic principles of design remain, and any outputs of the automated systems must be evaluated critically in terms of desirable practice.

Chapter 4

Example of a Rural Highway
Geometric Design

The preliminary geometric design of a rural highway described in this chapter illustrates the major elements of the design and cost estimating process and incorporates many of the considerations outlined in the earlier chapters.

The underlying philosophy in structuring the example design and specifying the required products in the projects of Chapter 5 is to involve the student in the process from a "blank map" situation to a finished report, yet not requiring excessive repetition or details of certain elements that are not needed to illustrate the overall design process. For example, the student is asked to design only one horizontal and one vertical curve in detail. Approximate methods such as using fewer cross sections in the earthworks estimation are suggested also, in the knowledge that in practical cases the appropriate level of detail will be employed for these computations.

EXAMPLE OF PRELIMINARY RURAL HIGHWAY DESIGN

Various elements of geometric highway layout are illustrated by examples in most texts on highway engineering (see bibliography, Chapter 1). Information about how the various elements are combined into a complete project (the main content of this book) may often be obtained from studies of specific routes. Usually these are available for inspection at state transportation and public works agencies.

The preliminary rural highway design illustrated in the following pages is presented in a format suitable for use in a student's project and may, of course, be varied as required. In addition, under the heading of "Notes," comments that help to explain or clarify the design and presentation have been added. These comments would not normally be included in preliminary design documents.

IMPORTANT NOTES:

The selection of the location for the proposed highway design shown on the example is entirely for illustrative purposes and does not imply any interest or intent to construct a highway at this location.

The preliminary design illustrated in this chapter should not be considered as a design appropriate for construction. Refinements, safety measures (in addition to those inherent in the design guidelines) where required, and other detailed investigations and engineering analyses will be required and are beyond the scope of this description.

OBJECTIVES OF EXAMPLE

As stated previously, the intent of these notes is to provide the opportunity for conducting the preliminary geometric design of a rural highway. This implies that only the essential features of the highway be specified in terms of:

- Topographic and cultural features of the environs of the highway, with appropriate map references.

- Beginning and end points of the highway.

- Purpose and need for highway.

- Anticipated traffic volumes.

- Design policy parameters (controls and designation).

- Highway authority having jurisdiction.

Thus, the project requires the full process of examining the terrain, sketching alternatives, making trial profiles, designing cross sections and drainage, and selecting one or more alternatives for preliminary design and costing. It is evident that there will be no unique solution to the problem, except when the terrain is of an extremely uniform configuration, and that each design must be judged upon its own merits. It is not the intent to necessarily design "the best" route in this instance, but one that conforms to the design designations and controls. The design typically would be modified later as a part of the overall project process.

SCOPE OF PRELIMINARY DESIGN

The scope of the preliminary design should be such that the resulting plans and specifications adequately inform individuals having an interest in the outcome of the project; engineers, right-of-way specialists, surveyors, economists, financial analysts, elected officials, environmentalists, and lay persons should be able to obtain from the preliminary engineering design documents the information they need to effectively contribute to the final design.

The following list, therefore, indicates the major items that, in most cases, should be presented in the preliminary engineering design documents:

- Title Page and Introductory Information

- Table of Contents

- Location Map

- Background and Specifications:

  - Introduction
  - Objectives
  - Design Policy
  - Design Designation
  - Design Controls
  - Lane Requirements
  - Other Conditions

- Major Environmental Features

- Alternative Routes

- Screening of Alternatives

- Horizontal Alignment

- Profile (Vertical Alignment) of Proposed Highway

- Coordination of Vertical and Horizontal Alignment

- Examples of Curve Design

- Cross Sections

- Earthworks Quantities

- Drainage:

    - Selected Catchment Area
    - Plan of Ditch, Culvert Layout
    - Runoff Computation and Check of Ditch Design

- Main Features of Intersection Design

- Construction Cost Estimate

- Summary of Likely Environmental Issues

- Summary of Key Technical Features

- Economic Analysis

- Environmental review (identification of key features only for future guidance)

- Requirements of various highway agencies will differ somewhat. However, the above list is reasonably representative and can be augmented or modified to suit specific needs.

TITLE PAGE

The following items and layout inform the readers about the major features:

PRELIMINARY DESIGN OF PROPOSED HIGHWAY

LOCATION:...

FOR:...

DESIGN AGENCY, FIRM, OR INDIVIDUAL:...

DATE:...

TABLE OF CONTENTS

This should clearly list each section and the relevant page numbers.

LOCATION MAP

The location map for the proposed highway is shown in Figure 4-1. It indicates the topographic features in the area of the route and the selected end points where connections with the existing roads are to be made. Obviously, the route cannot be shown on the location map until the preliminary design has been completed.

In addition, it is usually important that the location map be properly identified with regard to state, county, municipal, and other jurisdictional boundaries, particularly insofar as funding and technical details may be reviewed by different (including federal) agencies and other interests.   It is also usual to relate the proposed highway to the regional highway network.

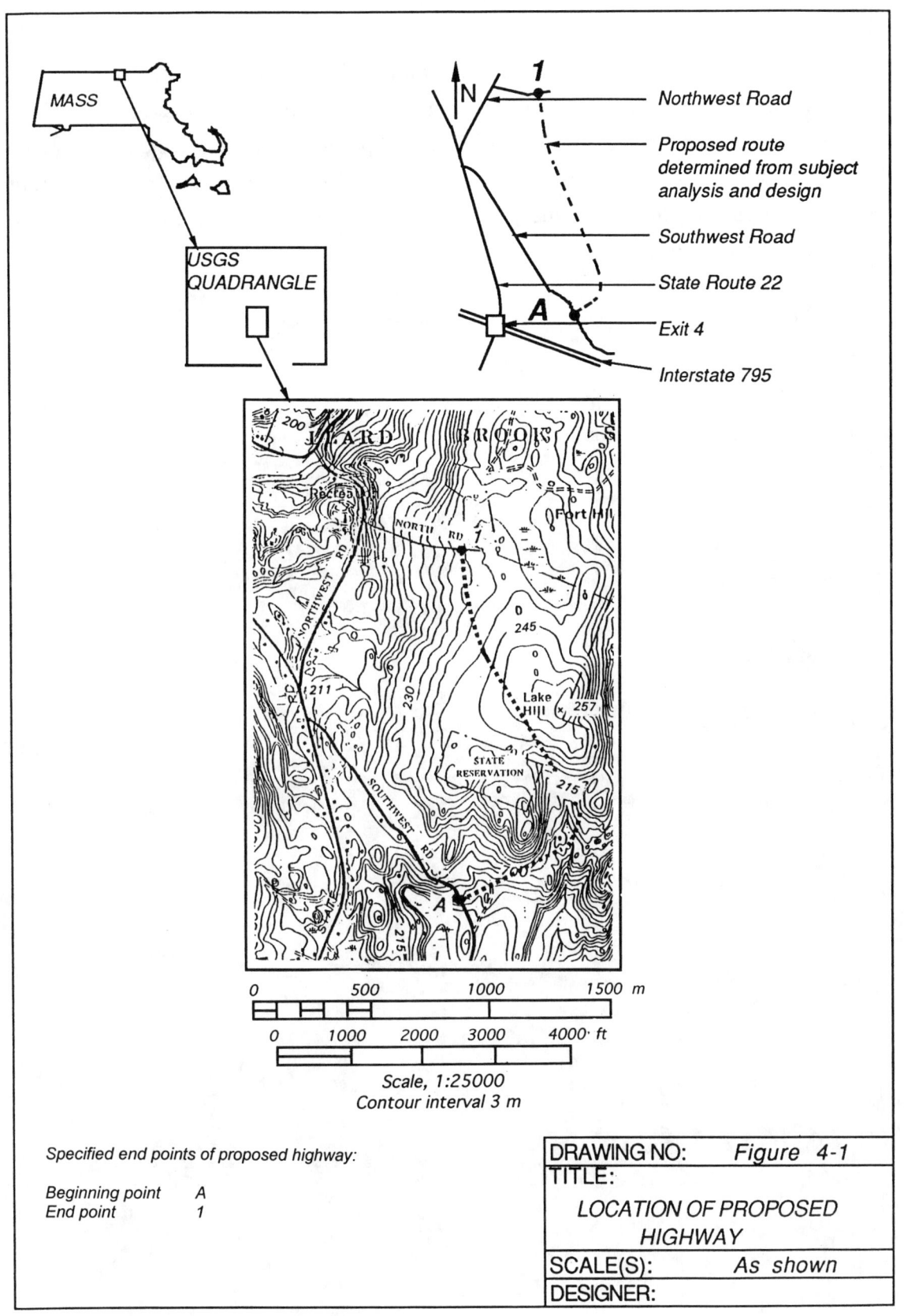

MASS

USGS QUADRANGLE

N

1

Northwest Road

Proposed route determined from subject analysis and design

Southwest Road

State Route 22

A

Exit 4

Interstate 795

| 0 | 500 | 1000 | 1500 | m |

| 0 | 1000 | 2000 | 3000 | 4000· | ft |

Scale, 1:25000
Contour interval 3 m

Specified end points of proposed highway:

Beginning point   A
End point   1

DRAWING NO:	Figure 4-1
TITLE:	
LOCATION OF PROPOSED HIGHWAY	
SCALE(S):	As shown
DESIGNER:	

BACKGROUND AND SPECIFICATIONS

This section provides the basic information that the designer relies upon to select the routes and undertake the design. The information should be clearly stated also because it informs others about the base guidelines and assumptions that have affected the outcome. These guidelines and assumptions may be the subject of future policy changes resulting from the review of the preliminary design but typically do not change during the preliminary design process.

In this project, the background and specifications are presented in Figures 4-2 and 4-3 and comprise:

- Introduction and background

- Objectives

- Design policy

- Design designation

- Design controls

- Lane requirements

- Other conditions

BACKGROUND AND SPECIFICATIONS --
PROPOSED CONNECTOR HIGHWAY

Introduction and Background:

The preliminary route selection and design of the proposed highway is intended to assist in exploring the costs and land - use implications of connecting North Road with Southwest Road at points A and 1, respectively, shown on the location map. Tentative analyses of future traffic levels have been made, including consideration of the potential use of the State Reservation as an environmental research area, with possible public access. Because of the steep grades in the area, the terrain is classified as 'mountainous'.

Objectives:

In accordance with the above background, it is required to conduct a preliminary route selection and design for a two-lane highway in accordance with AASHTO "rural collector" highway design standards, and estimate the capital, maintenance, and vehicle operating cost. An indication of key environmental concerns for more detailed analyses is also required. Two alternative routes are to be initially investigated. However, only one route is to be selected for the design process in this project, immediately following a brief screening of both routes.

Design Policy:

In addition to using the AASHTO policy on geometric design for a "rural collector" highway as the basic design guidelines, several more specific guidelines have been defined within the general policy, resulting from examination or the environs of the highway and its likely uses. Traffic control devices should be provided in accordance with the latest version of the MUTCD and local regulations.

Design Designation:

Terrain	= Mountainous
Classification	= Rural collector
Design Speed	= 60 km/h
ADT (Current Year, 1999)	= 0
ADT (Design Year, 2019)	= 3,500
K	= 10%
D	= 65%
DHV	= 350
T	= 2%

The design speed was selected based upon data in Table 2-4 where, for an ADT of over 2,000 in mountainous terrain, 60 km/h is the stated minimum design speed. This design speed was considered preferable to a higher value in order to provide greater flexibility in selecting a route through the constrictive terrain and the contiguous environmental features .

Design Controls

Design speed, V	= 60 km/h
Superelevation rate, e (max.)	= 6.0%
Minimum radius for above V and e	= 140 m → Center of Road
(Also satisfies stopping sight distance requirements for horizontal curves)	
Maximum grade, except at intersections (see below)	= 7%
Minimum grade for drainage purposes	= 0.5%
Maximum vertical curve K value, drainage criteria	= 51
Maximum fill height and cut depth at centerline	= 7 m

continued......

DRAWING NO:	Figure 4-2
TITLE: *BACKGROUND AND SPECIFICATIONS - 1*	
SCALE:	N/A
DESIGNER:	

Intersection Geometrics

The horizontal intersection of the proposed route with the existing highway should be a minimum of 30 m tangent segment as close to 90° as possible to the existing highway, but within the range of 75° to 105°. Maximum grade (positive or negative within 30 m of intersection) = 2%. Minimum grade at any location for drainage purposes = 0.5%.

Lane Requirements:

Two lanes are the minimum number or lanes for an adequate design, and a check with the tabulations of Chapter 2 indicates that a design speed of 60 km/h is consistent with level of service D to E for mountainous terrain. Therefore, the highway will comprise 2 - 3.6 m lanes with 2.4 m shoulders. These and other cross section characteristics are to be as indicated below, based upon the guidelines of Ref. 1.

Other Conditions:

In recognition of the preliminary nature of this route selection and design, of the mountainous nature of the terrain, the relatively short length of the proposed road, and the need to minimize environmental impacts, the following conditions shall apply:

1.   No bridges or tunnels are to be included in this preliminary design.
2.   Assume that the profile of the existing highway is equivalent to the adjacent contours.
3.   Climbing lanes are not required due to low traffic volumes.
4.   Superelevated segments of the highway may be detailed if desired, but are usually not a critical element of preliminary design.
5.   Assume 30% of excavation is in rock.
6.   Provision of overtaking distances is of minimal concern, due to the short length of the proposed road.
7.   Cross sections are to be designed as indicated earlier in Chapter 2 and shown graphically later in this chapter. More detailed requirements regarding slopes and guardrail also take into account the constricted space available for the proposed highway. The requirements are summarized as follows:

(a)   At all locations the surface from the centerline to the beginning of the fore slope shall comprise a 3.6 m pavement, 2.4 m shoulder, and a further 1.6 m. The 1.6 m distance will accommodate a guardrail if required and otherwise assist roadside safety and stability of the shoulder's edge.

(b)   In cut segments only, no guardrail is required. This is because, for a 60 km/h design speed, the required minimum clear zone distance is 3.3 m according to Figure 2-10, and the actual distance from the edge of the traveled way to the beginning of the fore slope is 4 m -- all at a slope of less than 10:1. The fore slope leading to the bottom of the ditch would then be 2:1, 1.3 m bottom and 2:1 ditch back slope. The side slope of the cutting would then be 1.5: 1.

(c)   On embankments only, where an embankment is 1 m high or more and a toe ditch is provided, or is 2 m high or more where no toe ditch is provided, a guardrail and a 2:1 fore slope shall be provided. Where greater heights exist than the 1 m and 2 m, above, a guardrail shall be provided and the embankment and ditch slopes shall be 1.75 to 1, with 1.3 m ditch bottoms, where applicable.

(d)   Where culverts and other obstructions may interfere with the passage of a vehicle that has unavoidably run off the road, a guardrail shall be provided.

(e)   All fore slopes shall be properly rounded where they intersect with the shoulder of the highway, but this need not be shown on the preliminary design drawings.

Important Note:

The dimensions and requirements described above are for the conditions associated with this particular project only. All design of actual elements must result from a detailed engineering and economic analysis performed by, or supervised by, a suitably qualified engineer.

DRAWING NO:	*Figure 4-3*
TITLE: *BACKGROUND AND SPECIFICATIONS - II*	
SCALE:	*N/A*
DESIGNER:	

MAJOR ENVIRONMENTAL FEATURES

The major topographic, cultural, soil, and vegetation features are identified in Figure 4-4. Knowledge of the location of these features enables the routes to be laid out in a way that is sensitive to the various features. For this project we would wish to avoid swamps (suggested minimal clearance, 60 m from edge of right-of-way), the state reservation and, if possible, grades steeper than 10%. Identification of streams and potential watercourses as indicated by contour features, is important to assist in the location of culverts that will be required in the drainage design described later.

Note that a more detailed, but separate, environmental study would probably be carried out along with this preliminary highway design. Such a study, however, cannot be fully completed until a route is selected and a preliminary design is proposed.

ALTERNATIVE ROUTES

From the examination of the contours and terrain, and using the approach described in Chapter 3, two possible routes, A and B, were defined, as shown in Figure 4-5. In the plan view of these routes, the curves have been shown to conform with the minimum required radius. In the profiles shown in Figure 4-6, both alternatives comply with the required controls, and Route B is approximately 230 m longer than A.

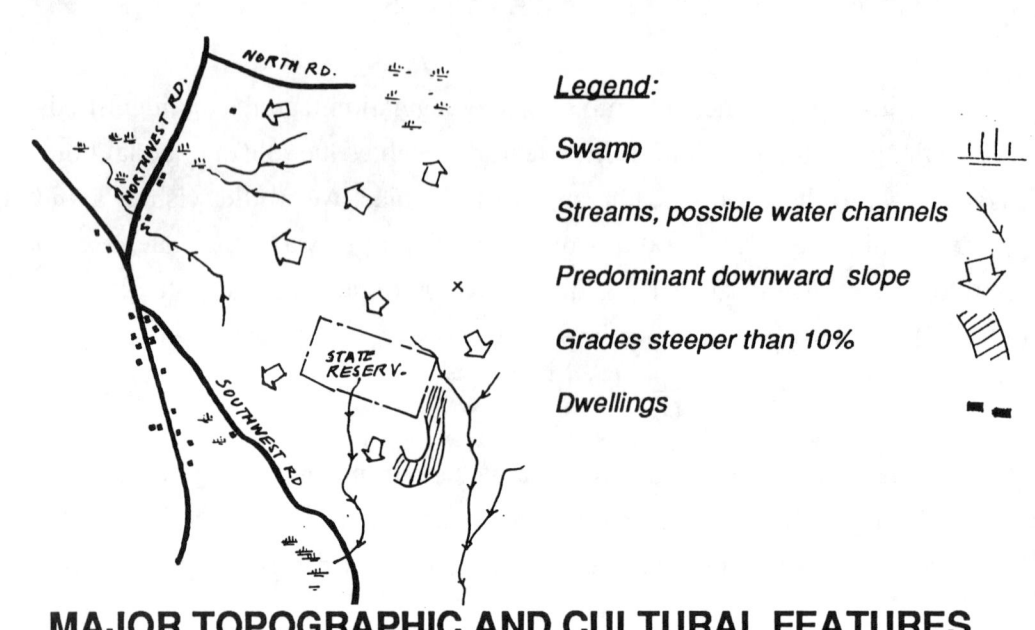

Legend:

Swamp

Streams, possible water channels

Predominant downward slope

Grades steeper than 10%

Dwellings

MAJOR TOPOGRAPHIC AND CULTURAL FEATURES

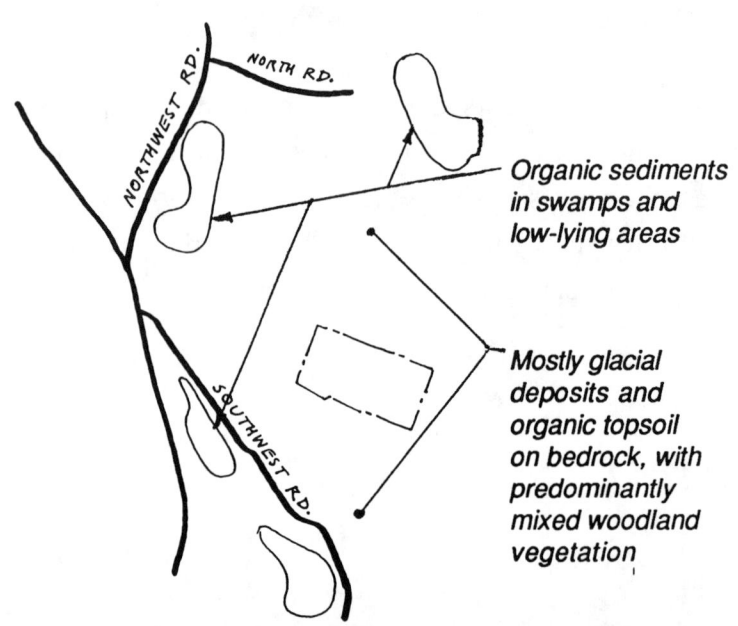

Organic sediments in swamps and low-lying areas

Mostly glacial deposits and organic topsoil on bedrock, with predominantly mixed woodland vegetation

PREDOMINANT SOIL AND VEGETATION CHARACTERISTICS

Source: Overlays of Figure 1 - 1

DRAWING NO:	Figure 4-4
TITLE: CONTIGUOUS ENVIRONMENTAL FEATURES	
SCALE(S):	As shown
DESIGNER:	

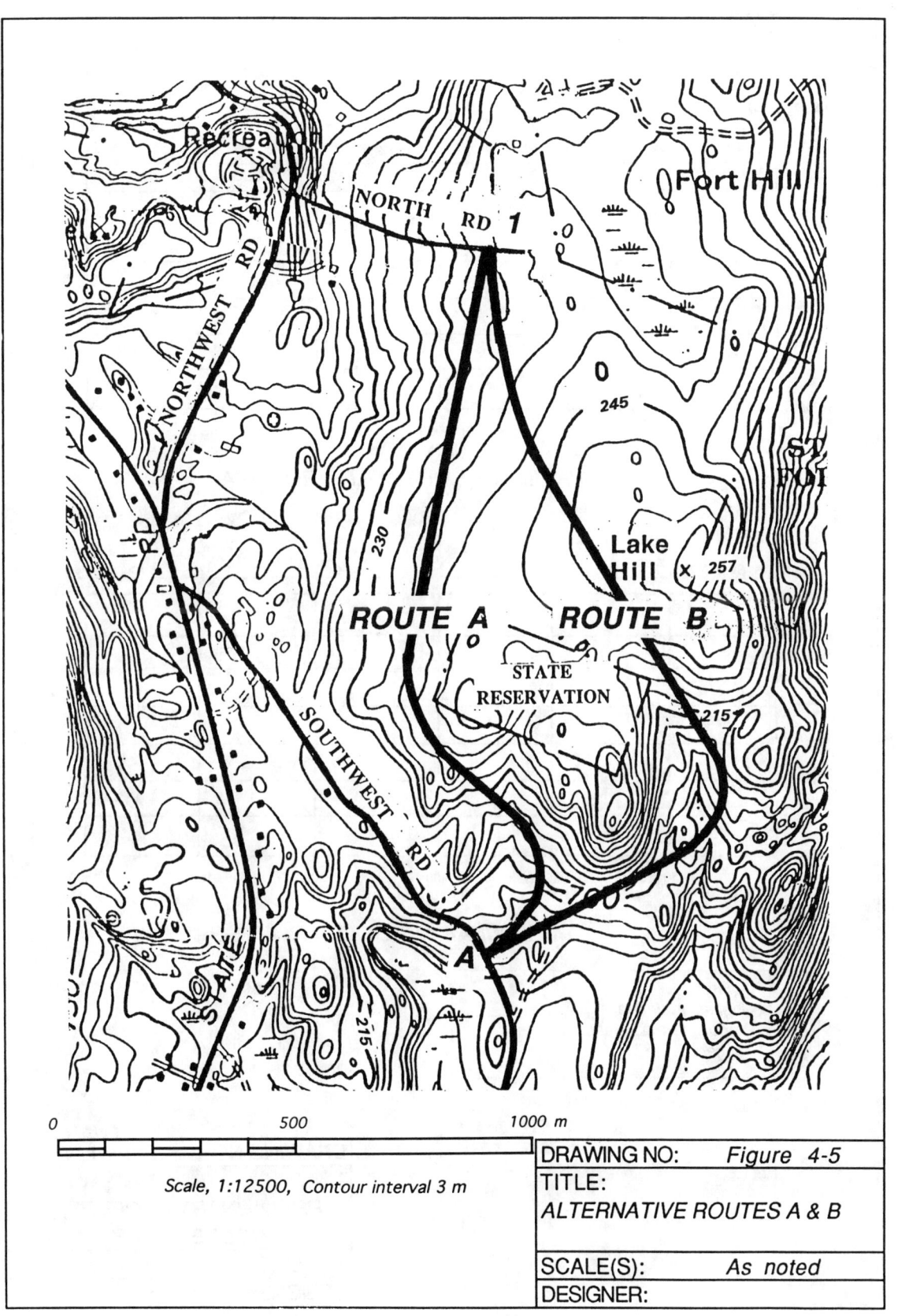

Elevation, m

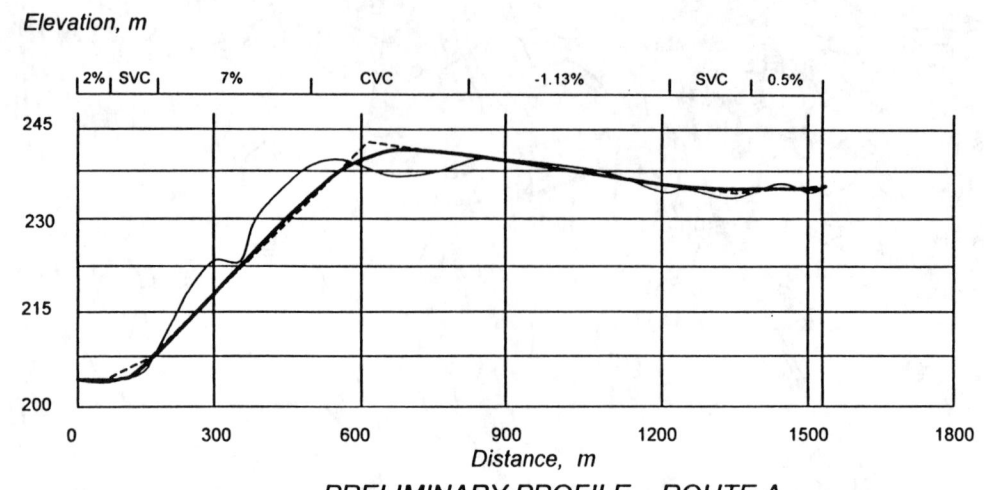

PRELIMINARY PROFILE – ROUTE A

Elevation, m

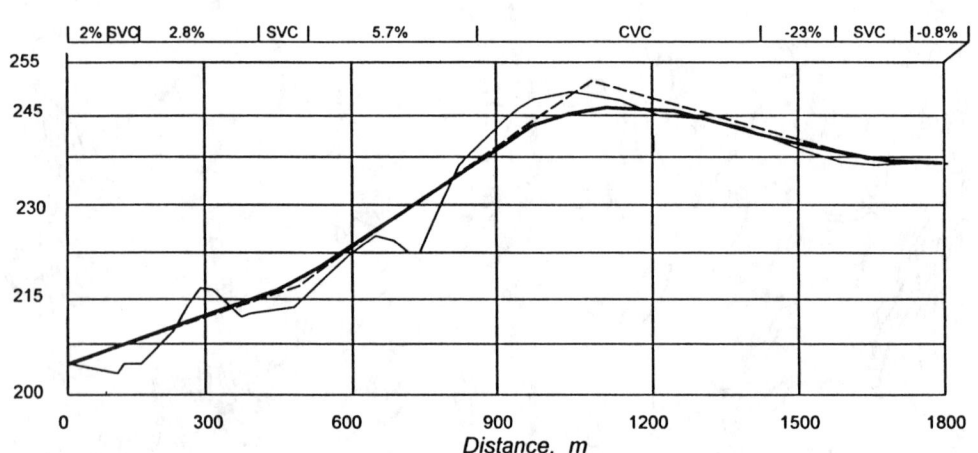

PRELIMINARY PROFILE – ROUTE B

Legend:
Original ground level _____
Proposed pavement centerline ▬▬▬
Tangent - - - - -

DRAWING NO:	Figure 4-6
TITLE:	
PRELIMINARY PROFILES, ROUTES A & B	
SCALE(S):	As shown
DESIGNER:	

SCREENING OF ALTERNATIVES

Examination of the profiles of Routes A and B, shown in Figure 4-7, indicates that both Routes A and B could provide a vertical profile that would conform with the gradient and depth of cut and fill requirements, but that Route A would require excessive uphill haul. For this and other reasons indicated in the summary screening process shown in Figure 4-7, Route B is the one selected for technical feasibility and cost estimates covered in the succeeding sections.

Note: Depending upon the circumstances and location of a proposed design, other criteria than those listed might be applied in the screening process. For example, right-of-way, soil conditions, and/or environmental concerns might have required greater consideration. Also, because there may not have been a significant difference in likely merit discerned at this stage between alternatives, perhaps two or more alternatives might have been investigated prior to the preliminary design stage, instead of only one alternative.

HORIZONTAL ALIGNMENT

Based on the methods described in Chapter 3, the horizontal alignment can be developed for alternative B and the relevant dimensions and angles computed and checked. The resulting traverse is shown in Figure 4-8. The development of tangent and curve distances may then be converted to stations and an estimate of the traverse closure made to ensure that no gross errors have been made in the scaling of distances and angles. These computations are tabulated in Figure 4-9. No adjustment of the traverse is shown here, but this could be done if required.

In addition to the dimensions of each of the tangent and curve segments of the proposed highway, the stations are computed for points along the centerline of the route.

Care must be taken to define clearly the location of the beginning and end points of the proposed route, in this case assumed to be the centerlines of the existing highways. Each curve radius is checked for compliance with the minimum allowable radius of 135 m.

CRITERIA	SCREENING EVALUATION	
	ROUTE A	ROUTE B
1. Length of route	1590 m	1800 m
2. Conformance with design controls	Full conformance	Full conformance
3. Cut and fill balance	Excess of fill due to deep cuts necessitated by 7% grade limit	Balance appears acceptable
4. Need for bridges or special structures	None	None
5. Environmental impacts	Need to haul and dispose of excess fill may cause environmental problems	No obviously excessive environmental impacts
6. Potential high cost items	Excessive hauling and disposal away from site	Drainage precautions may be needed due to wetlands west and downhill of route
7. Haul direction	Excessive uphill haul	Appears satisfactory
8. Other	–	–
9. Other	–	–

CONCLUSIONS: Route A is shorter than Route B. However, the excessive uphill haul and need to dispose of considerable soil off the site for Route A could result in high costs.
Decide to proceed with Route B for analysis purposes.

DRAWING NO:	Figure 4-7
TITLE: SCREENING ALTERNATIVES	
SCALE(S):	N/A
DESIGNER:	

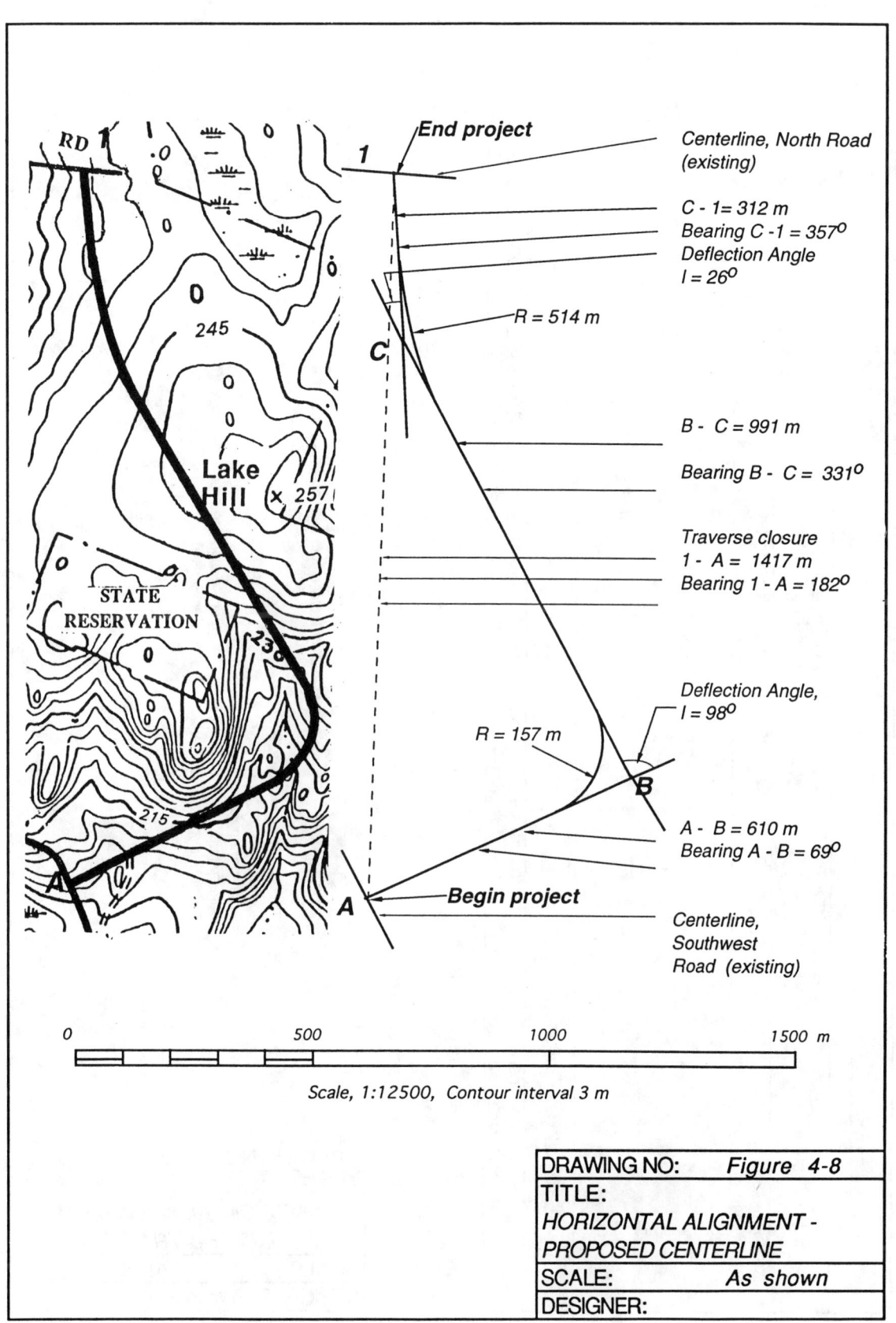

Scale, 1:12500, Contour interval 3 m

DRAWING NO:	Figure 4-8
TITLE:	
HORIZONTAL ALIGNMENT - PROPOSED CENTERLINE	
SCALE:	As shown
DESIGNER:	

DEFINING THE TRAVERSE AND ROUTE:

REF.	PI DISTANCE	I	RADIUS R	DEGREE of CURVE	SUBTANT T	ARC L	TANGENT LENGTH	CUMULATIVE DISTANCES		
								PC	PT	END
(From route diagram)	(Scaled)	(Scaled, degrees)	(Scaled)	$\frac{1746.00}{R}$	(R*Tanl/2)	(R*Irad)	(PI dist.- Subtan'ts Length)	(PT+Tg(PC+Arc Length) Length)		(PT+Tgt Length)
A		0.00								
A-B	610				0		429			0
B		98.00	157	11.12	181	269		429	429	0
B-C	991				0		691		697	0
C		26.00	514	3.40	119	233		1389	1622	0
C-D	312				0		194			1816

1815.55

TRAVERSE CHECK

REF	PI DISTANCE	DEFLECT ANGLE I	CUMUL I	QUAD	ANGLE	NORTHEAST		SOUTHEAST		SOUTHWEST		NORTHWEST		SUMMARY	
						LAT	DEP	LAT	DEP	LAT	DEP	LAT	DEP	LAT	DEP
A-B	610	69.00	69.00	1	69.00	219	569	0	0	0	0	0	0	219	569
B-C	991	-98.00	-29.00	4	29.00	0	0	0	0	0	0	867	-480	867	-480
C-D	312	26.00	-3.00	4	3.00	0	0	0	0	0	0	312	-16	312	-16
D-A	1417	-355.00	-358.00	3	2.00	0	0	0	0	-1416	-49	0	0	-1416	-49

Traverse closure errors: -19 23

CONCLUSION: Error is less than 2% of total length, and no gross error due to scaling is evident

Distances are in meters
Angles are in degrees
Right turn angles are positive
Latitude is COS(CUMUL I) for North or South, depending on the quadrant
Similarly, Departure is SIN(CUMUL I) for East or West

DRAWING NO:	Figure 4-9
TITLE:	HORIZONTAL ALIGNMENT SUMMARY & TRAVERSE
SCALE:	As shown
DESIGNER:	

Note: It may be considered necessary in practice to re-compute the angles and distances of the traverse if closure is not obtained. However, in this exercise where angles and distances were scaled, the accuracy is considered sufficient for the preliminary geometric design if the error of closure is within about 5% of the total length of the highway. A re-computation would typically be performed either prior to or in conjunction with a field survey.

During this preliminary design stage for this class of highway, no attempt has been made to use transition curves or to detail the attainment of superelevation because this is unlikely to significantly affect the general route or the construction cost If the designer wishes, however, the superelevation and transition curves could be calculated and shown on the drawings.

PROFILE (VERTICAL ALIGNMENT)

Based upon the horizontal alignment established earlier and incorporating any modifications to it, the profile at the centerline of the proposed highway is plotted as shown in Figure 4-10. The location and lengths of all vertical curves and of positive and negative gradients are clearly shown. All the controls (maximum grade, maximum and minimum length of curve, maximum vertical cut and fill dimensions, and approximate balance of cut and fill) have been complied with. The following summarizes the curve selection process:

1. The minimum length of the curve is based upon stopping sight distance criterion.

2. The maximum length of the curve for this project is based upon the maximum K value established for the drainage criteria. In Chapter 2, the maximum value of K for drainage purposes is 51. Often, other agencies may have their own values. Note that the drainage criteria do not necessarily indicate that the slope will not drain, only that special attention should be given to the drainage design.

3. Select a curve length that is equal to or greater than the minimum length for stopping sight distance criteria and equal to or less than the maximum for the drainage criteria.

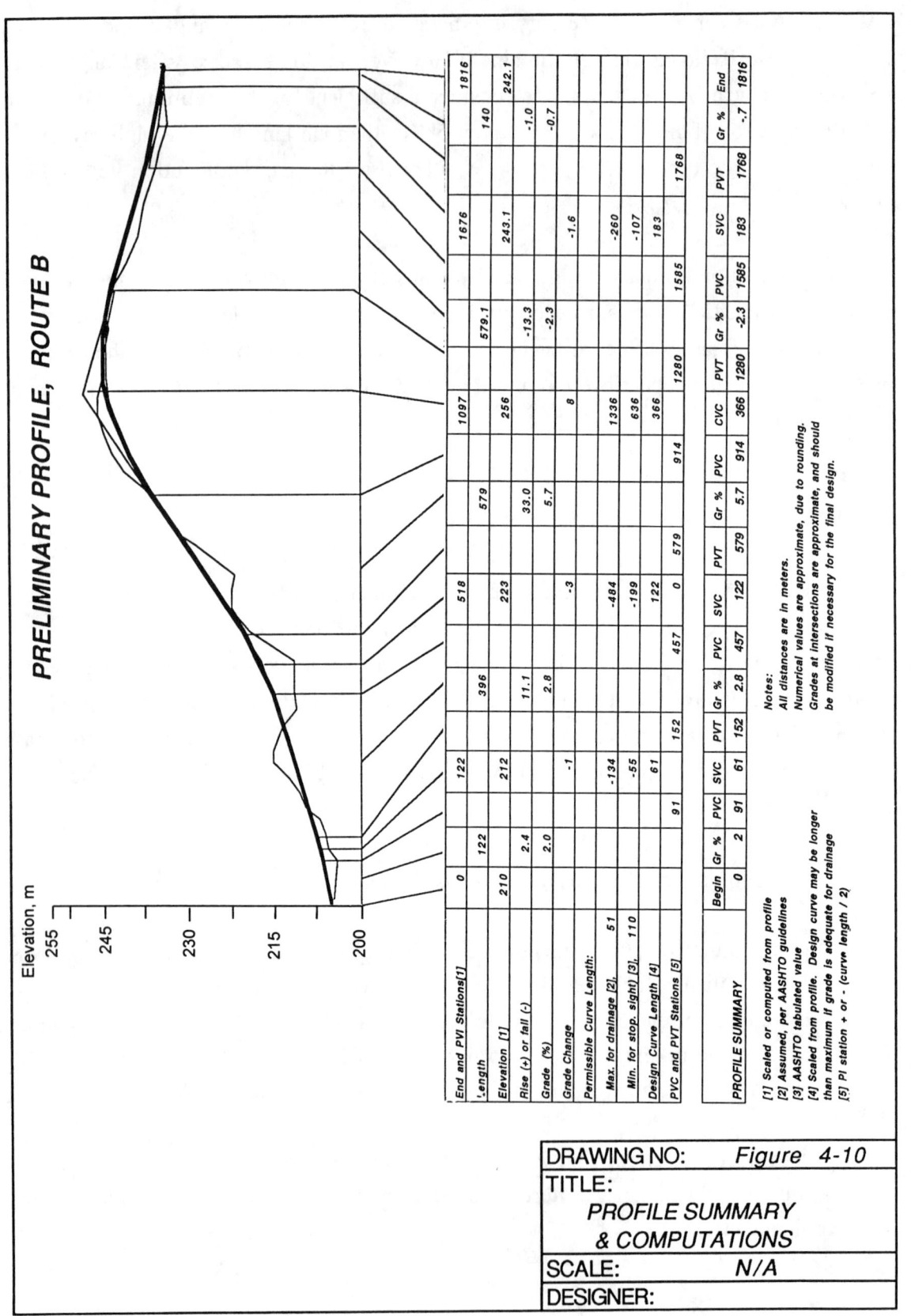

PRELIMINARY PROFILE, ROUTE B

DRAWING NO:	Figure 4-10
TITLE:	PROFILE SUMMARY & COMPUTATIONS
SCALE:	N/A
DESIGNER:	

4. If the selected length of curve based upon the above consideration is not acceptable when the overall profile is considered (e.g., results in excessive cut or fill, or interferes with an adjacent curve), the profile must be re-examined and adjusted, and the process repeated until all criteria are met.

EXAMPLES OF CURVE DESIGNS

Figures 4-11 and 4-12 show the computations for selected horizontal and vertical curves, respectively, to provide greater detail and assist possible staking for a more extensive field survey.

COORDINATION OF HORIZONTAL AND VERTICAL ALIGNMENTS

A graphical comparison is shown in Figure 4-13, together with comments alongside the criteria presented in Chapter 2. Examination of the criteria indicates that no significant problems exist with the proposed route, although some refinement may be necessary in a final design.

CROSS SECTIONS

The cross sections are based upon the selected design elements shown in Figure 4-14 that are consistent with the cross section design dimensions described in the background and specifications (Figures 4-2 and 4-3). See the Appendix for notes on constructing the cross sections. Note that the cross section elevations for earthmoving are not the same as the pavement elevations because the pavement thickness must be allowed for. For this highway, we are assuming a total pavement thickness (surface, base, and subbase) of 0.5 m, as shown in the Appendices. Figures 4-15 through 4-18 show the cross sections for earthwork computations.

Note that in this example, cross sections have been taken at approximately 150 m intervals instead of the more usual intervals of between about 10 m and 30 m, depending on the terrain. The pavement, shoulder, slope, and drainage details are in accordance with the earlier discussions of these topics. The 150 m intervals have been used to illustrate the process only. In certain instances of preliminary design, this will provide an indication of the extent of the balance between cut and fill but may not always provide an adequate indication of haul lengths. This latter consideration is not investigated further in this project.

EXAMPLE OF CIRCULAR CURVE COMPUTATIONS
Curve Ref. B.

1. INPUT FROM CENTERLINE OF SELECTED CURVE

Station PI	610.00
Tangent deflection angle	98.00
Radius	157.00

2. CURVE COMPUTATIONS:

(A) Summary of curve dimensions:

Degree of curvature, Da	11.12	[Da = 1746.40 / R]
Radius, R	157.00	[R = 1746.40 / Da]
Tangent distance, T	180.61	[T = R*TAN(I/2)]
The external, E	82.31	[E = R*(1/COS(I/2)-1)]
Mid-ordinate, M	54.00	[M = R*(1-COS(I/2))]
Long chord, LC	236.98	[LC = 2*R*SIN(I/2)]
Curve (arc) length, L	268.53	[L = I/Da*100]
Station PC	429.39	[PC = PI - T]
Station PT	697.92	[PT = PC + L]

(B) Deflection angles along curve at 25 m intervals:

STATION on CURVE	POINT	CUMULATIVE STA. CHORD (ARC)	DEFLECT-ION ANGLE (0.5*L/Da)	ANGLE INCREMENT
429.39	PC			
450.00		20.61	3.76	3.76
475.00		25.00	8.32	4.56
500.00		25.00	12.88	4.56
525.00		25.00	17.45	4.56
550.00		25.00	22.01	4.56
575.00		25.00	26.57	4.56
600.00		25.00	31.13	4.56
625.00		25.00	35.69	4.56
650.00		25.00	40.26	4.56
675.00		25.00	44.82	4.56
697.92	PT	22.92	49.00	4.18

Check: Cumulative station deflection angle at PT = 2 x Deflection angle

2 x 49 = 98, therefore OK

DRAWING NO:	Figure 4-11
TITLE: *EXAMPLE OF HORIZONTAL CURVE COMPUTATIONS*	
SCALE(S):	*As shown*
DESIGNER:	

EXAMPLE OF VERTICAL CURVE COMPUTATIONS

Curve Ref: *Sag PVI = 1676 m (Figure 4-10)*

1. INPUT FROM CENTERLINE OF SELECTED CURVE

ITEM	SOURCE	VALUE	
G1	Profile	-2.30	
G2	Profile	-0.70	
PVC	Profile	1585.00	
PVT	Profile	1768.00	
L	Profile	183.00	
PVI	PVC+L/2	1676.50	
A	G1-G2	-1.60	
E	LA/800	-0.37	
PVI El	Profile	243.10	
PVC El	PVI-G1/100*L	245.20	
PVT El	PVI+G2/100*	242.46	
X: Low point	L*G1/A	183.00	*(Low point at PVT, due to adjacent -ve grades)*
X: Station	PVC+X	1768.00	

2. COMPUTED CURVE CENTERLINE ELEVATIONS AT 25 m INTERVALS

STATION	x	TGT ELN (PVC El+ x * G1/100)	y (4E* (x/L)SQU))	CURV ELN (Tgt El - y)	
1585.00	0.00	245.20	0.00	245.20	PVC
1610.00	25.00	244.63	-0.03	244.66	
1635.00	50.00	244.05	-0.11	244.16	
1660.00	75.00	243.48	-0.25	243.73	
1685.00	100.00	242.90	-0.44	243.34	
1710.00	125.00	242.33	-0.68	243.01	
1735.00	150.00	241.75	-0.98	242.74	
1760.00	175.00	241.18	-1.34	242.52	
1768.00	183.00	241.00	-1.46	242.46	PVT

(Checks with input PVT el'n, above)

DRAWING NO:	Figure 4-12
TITLE: *EXAMPLE OF VERTICAL CURVE COMPUTATIONS*	
SCALE(S):	N/A
DESIGNER:	

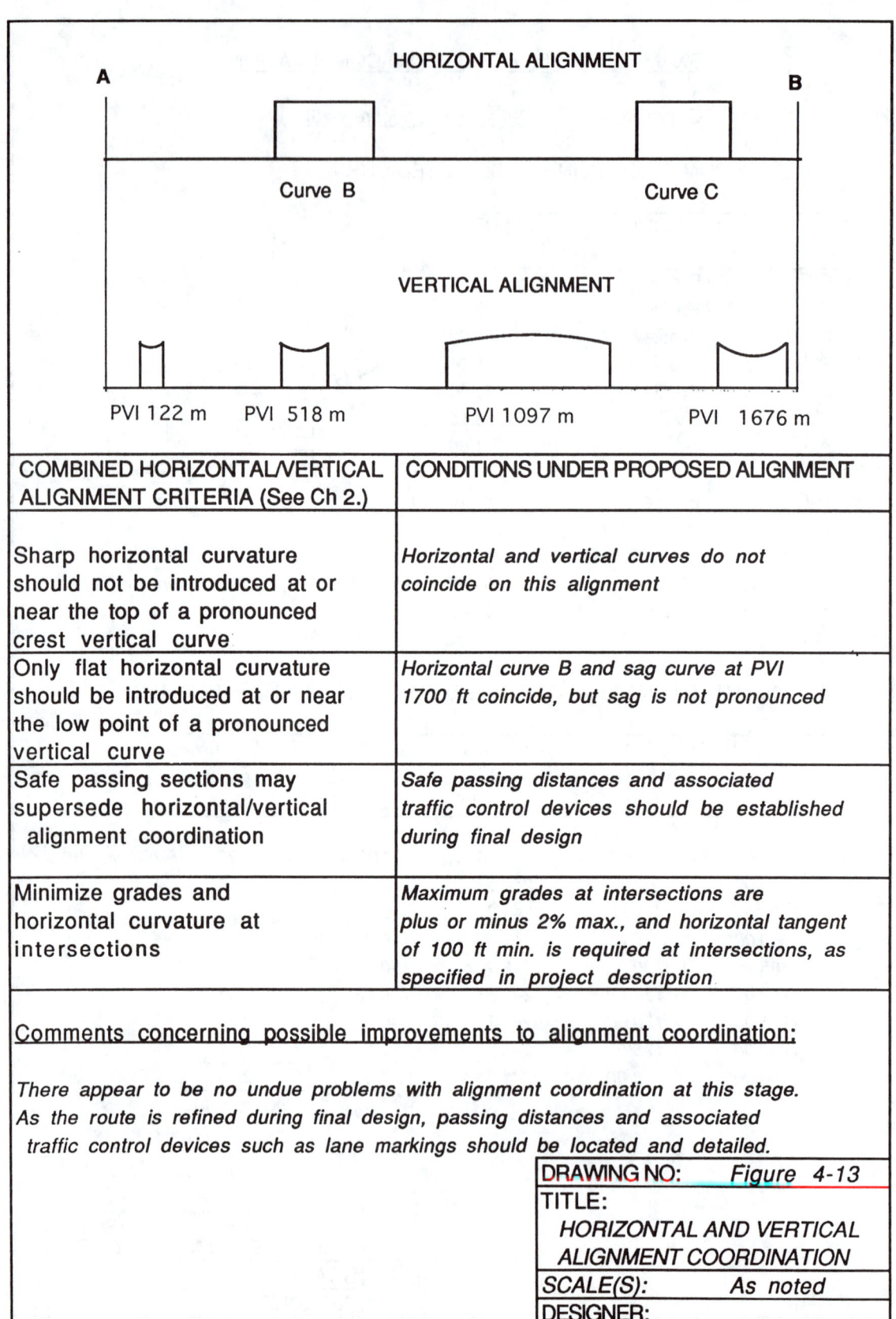

HORIZONTAL ALIGNMENT

A   B

Curve B   Curve C

VERTICAL ALIGNMENT

PVI 122 m   PVI 518 m   PVI 1097 m   PVI 1676 m

COMBINED HORIZONTAL/VERTICAL ALIGNMENT CRITERIA (See Ch 2.)	CONDITIONS UNDER PROPOSED ALIGNMENT
Sharp horizontal curvature should not be introduced at or near the top of a pronounced crest vertical curve	*Horizontal and vertical curves do not coincide on this alignment*
Only flat horizontal curvature should be introduced at or near the low point of a pronounced vertical curve	*Horizontal curve B and sag curve at PVI 1700 ft coincide, but sag is not pronounced*
Safe passing sections may supersede horizontal/vertical alignment coordination	*Safe passing distances and associated traffic control devices should be established during final design*
Minimize grades and horizontal curvature at intersections	*Maximum grades at intersections are plus or minus 2% max., and horizontal tangent of 100 ft min. is required at intersections, as specified in project description*

Comments concerning possible improvements to alignment coordination:

There appear to be no undue problems with alignment coordination at this stage.
As the route is refined during final design, passing distances and associated
traffic control devices such as lane markings should be located and detailed.

DRAWING NO:	*Figure 4-13*
TITLE:	*HORIZONTAL AND VERTICAL ALIGNMENT COORDINATION*
SCALE(S):	*As noted*
DESIGNER:	

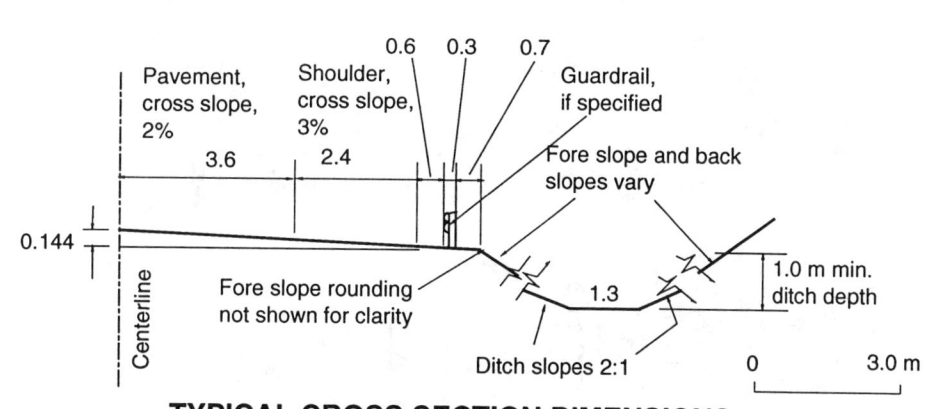

TYPICAL CROSS SECTION DIMENSIONS

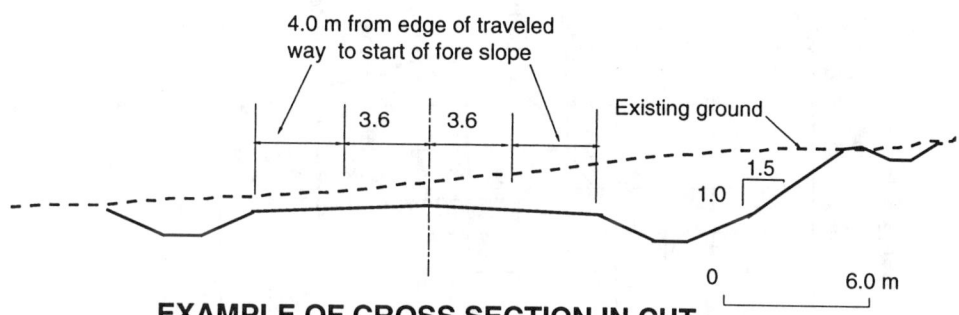

EXAMPLE OF CROSS SECTION IN CUT

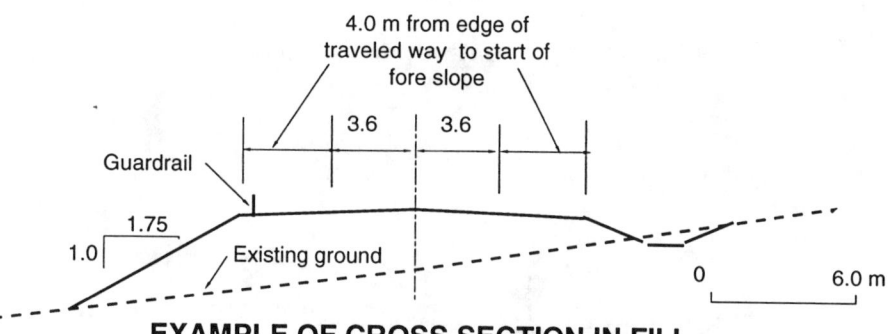

EXAMPLE OF CROSS SECTION IN FILL

All dimensions are in meters

Notes: The dimensions and configurations of the cross sections shown here are based upon the background and specifications described earlier, and are intended for use in this illustrative project only. Detailed analysis of cross sections and roadside features, and the resulting designs, would be required as a basis for the final design and for other projects.

DRAWING NO:	Figure 4-14
TITLE: TYPICAL CROSS SECTIONS FOR PROJECT	
SCALE(S): As shown	
DESIGNER:	

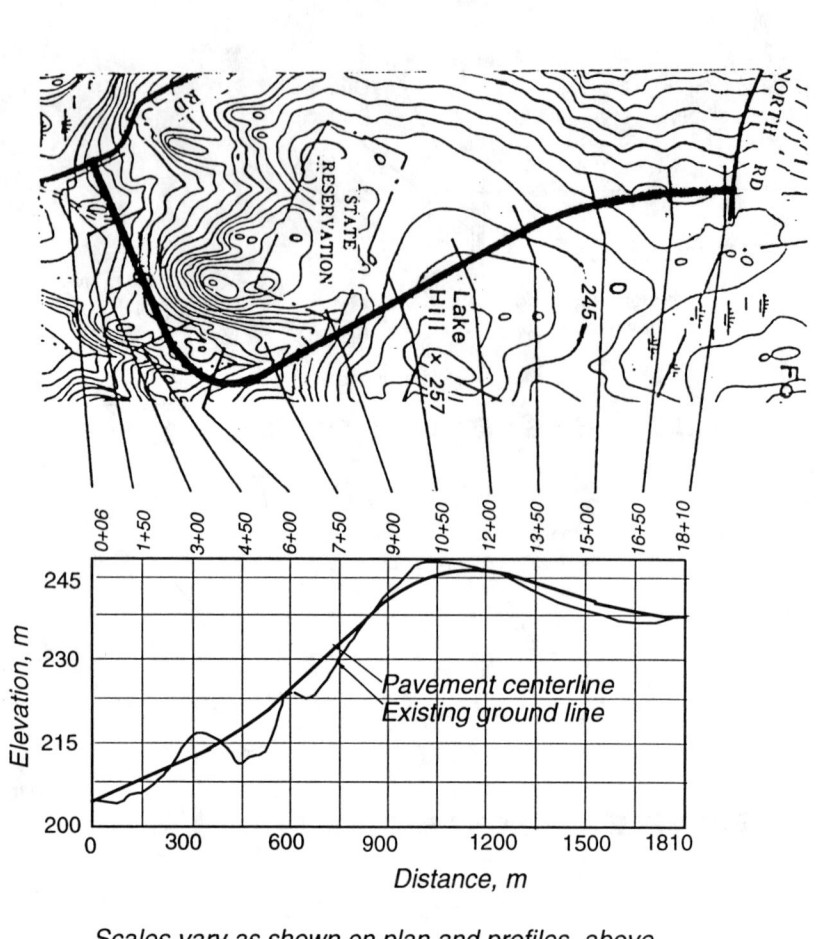

Scales vary as shown on plan and profiles, above

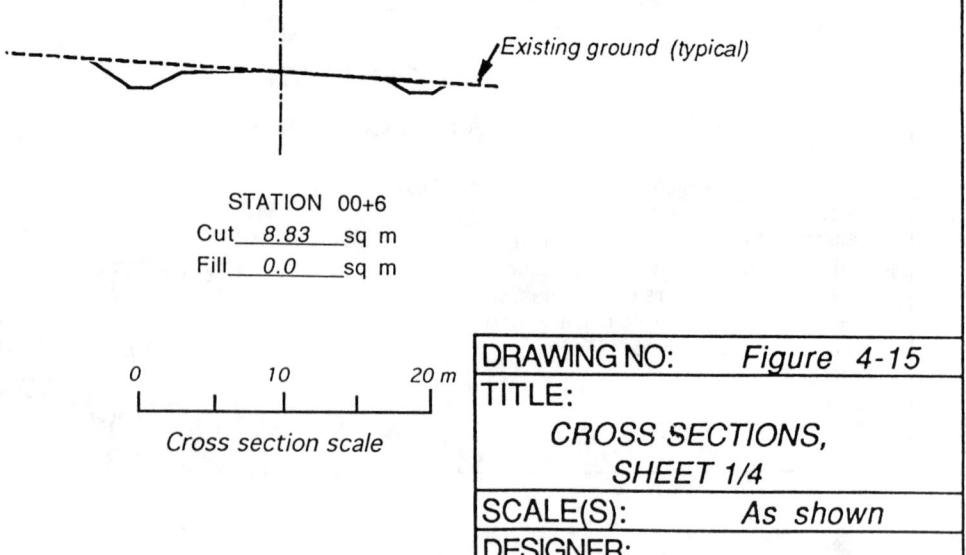

STATION 00+6

Cut ___8.83___ sq m

Fill ___0.0___ sq m

Cross section scale

DRAWING NO:	*Figure 4-15*
TITLE:	
	CROSS SECTIONS,
	SHEET 1/4
SCALE(S):	*As shown*
DESIGNER:	

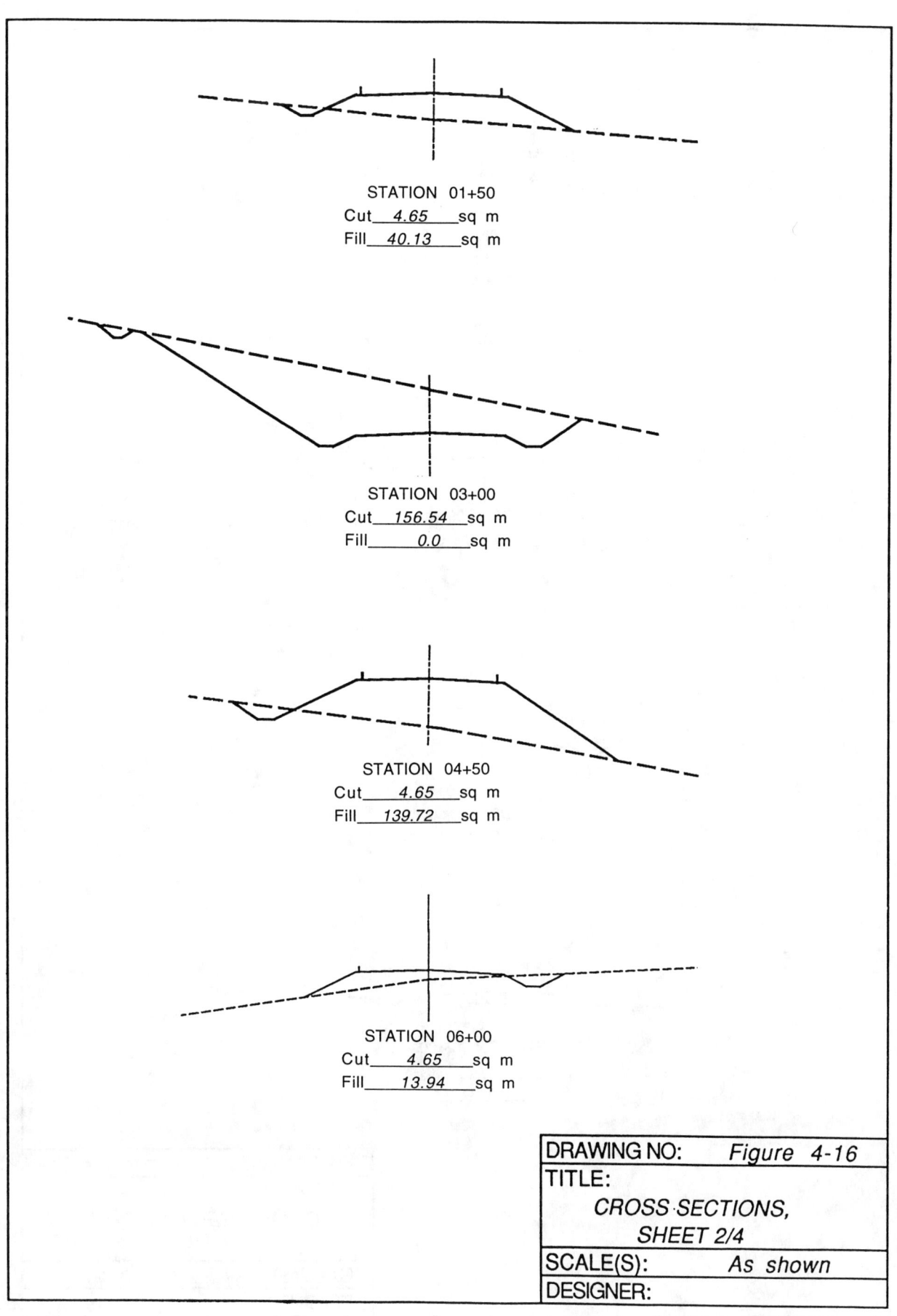

STATION 01+50
Cut ___4.65___ sq m
Fill ___40.13___ sq m

STATION 03+00
Cut ___156.54___ sq m
Fill ___0.0___ sq m

STATION 04+50
Cut ___4.65___ sq m
Fill ___139.72___ sq m

STATION 06+00
Cut ___4.65___ sq m
Fill ___13.94___ sq m

DRAWING NO:	Figure 4-16
TITLE: CROSS·SECTIONS, SHEET 2/4	
SCALE(S):	As shown
DESIGNER:	

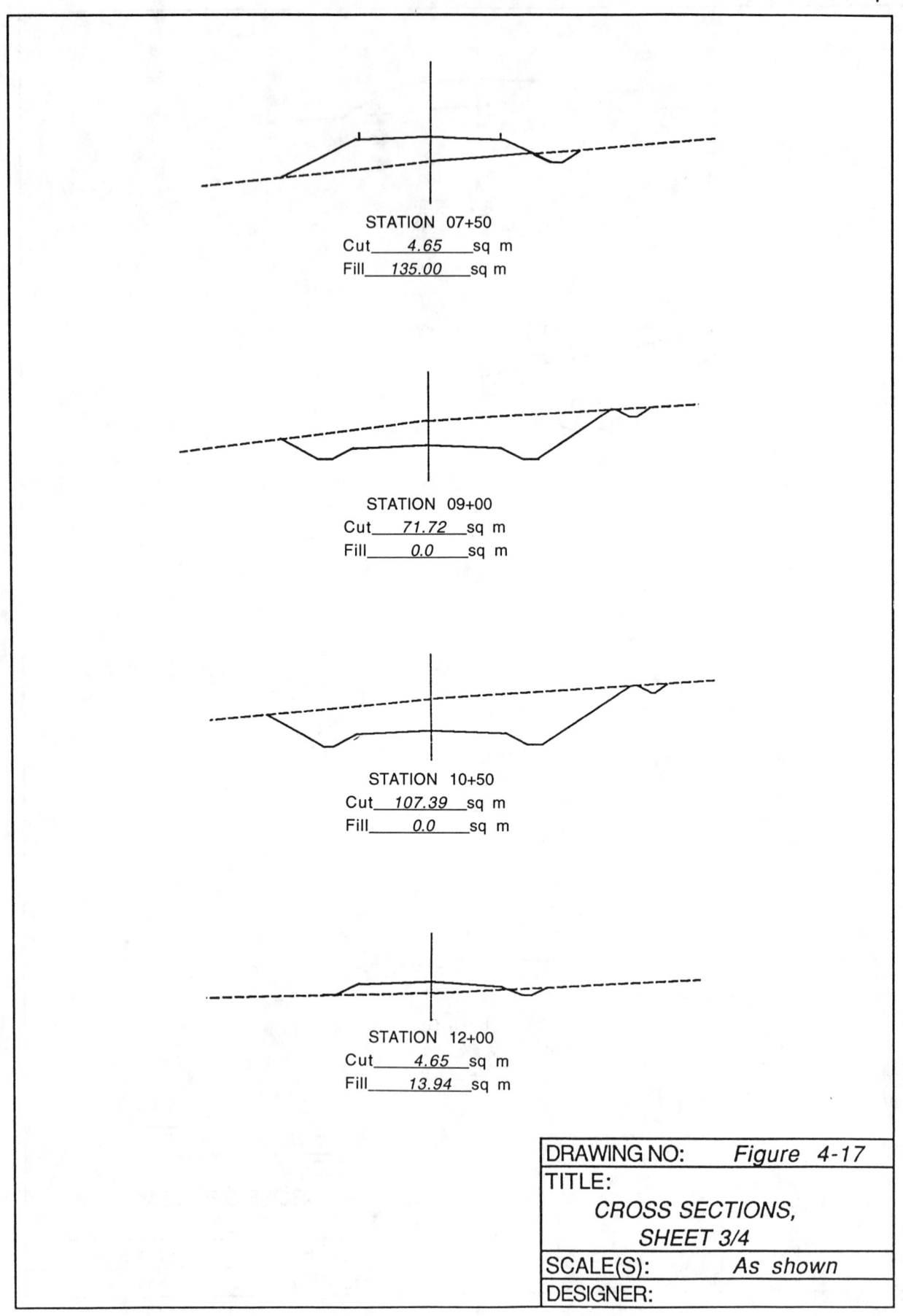

STATION 07+50
Cut ___4.65___ sq m
Fill ___135.00___ sq m

STATION 09+00
Cut ___71.72___ sq m
Fill ___0.0___ sq m

STATION 10+50
Cut ___107.39___ sq m
Fill ___0.0___ sq m

STATION 12+00
Cut ___4.65___ sq m
Fill ___13.94___ sq m

DRAWING NO:	Figure 4-17
TITLE:	
	CROSS SECTIONS, SHEET 3/4
SCALE(S):	As shown
DESIGNER:	

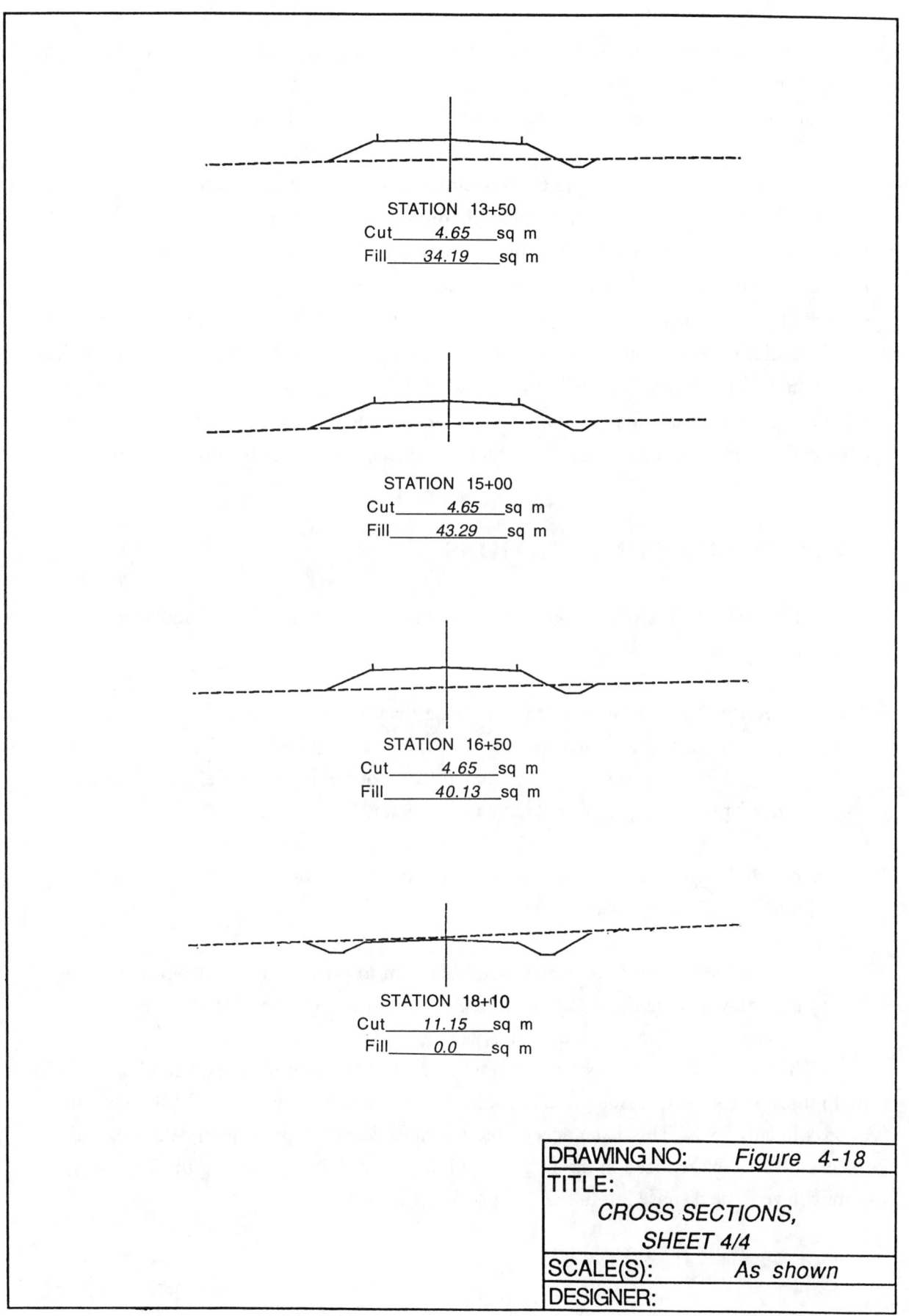

STATION 13+50

Cut____4.65____sq m

Fill____34.19____sq m

STATION 15+00

Cut____4.65____sq m

Fill____43.29____sq m

STATION 16+50

Cut____4.65____sq m

Fill____40.13____sq m

STATION 18+10

Cut____11.15____sq m

Fill____0.0____sq m

DRAWING NO:	*Figure 4-18*
TITLE:	
CROSS SECTIONS,	
SHEET 4/4	
SCALE(S):	*As shown*
DESIGNER:	

Depending upon the time available for the project, students should reduce the intervals to less than 150 m in order to increase the accuracy. Special sections may also be taken at critical areas where walls, deep cuts, or large fills would apply.

A visual examination of the cross sections indicates that in several cases the height of embankment slopes for cuts and fills where the route traverses a steep hillside can be considerably more than the depth at the centerline and that the width of the right-of-way increases accordingly. This is because the depth of cut or height of fill increases as the distance from the centerline increases. In these cases, the heights or depths may become considerable, causing visual scars in the landscape and requiring excessive right-of-way acquisition. The alternatives include reconsidering the alignments to reduce the amounts of cut and fill, introducing retaining walls, increasing the slopes based upon detailed engineering analysis, and using usually more expensive slope stabilization techniques.

EARTHWORKS COMPUTATIONS

The computation of cut and fill and mass-haul volume is conducted at this preliminary stage for the following purposes:

1. To verify that an approximate balance has been achieved in the amount of excavation (cut) and embankment (fill) material required for the route alignment selected. In cases where the cut and fill cannot be approximately balanced, some borrow or spreading areas will have to be obtained.

2. To provide an estimate of earthmoving quantities that will be needed for the preliminary cost estimate.

3. To provide the basis for a mass-haul diagram to provide a visual representation of the earthwork quantities and as a guide to key aspects of construction planning.

The cut and fill and mass-haul computations and diagram for the example are shown in Figure 4-19. The result of the computations shows an excess of fill over cut of approximately 5%. This is an acceptable tolerance for this preliminary design, and the excess fill may be reduced in the final design, if required, by modifying the alignment; or it may have to be disposed of in the vicinity of the project.

STATION (XX+XX) Distance	AREA OF CROSS SECTION EXC (Sq m)	EMB (Sq m)	EMB.+10	AVE AREA EXC	AVE AREA EMB	VOLUME EXC (Cu m)	VOLUME EMB	EXCESS (+ EMB) (- EXC)	MASS CURVE ORDINATE
6	10	0	0						
				7.5	22	1080	3168	2088	2088
150	5	40	44						
				80.5	22	12075	3300	-8775	-6687
300	156	0	0						
				80.5	77	12075	11550	-525	-7212
450	5	140	154						
				5	84.7	750	12705	11955	4743
600	5	14	15.4						
				5	32.5	750	4868	4117.5	8860.5
750	5	45	49.5						
				38.5	24.8	5775	3713	-2062.5	6798
900	72	0	0						
				89.5	0	13425	0	-13425	-6627
1050	107	0	0						
				56	7.7	8400	1155	-7245	-13872
1200	5	14	15.4						
				5	26.4	750	3960	3210	-10662
1350	5	34	37.4						
				5	42.4	750	6353	5602.5	-5059.5
1500	5	43	47.3						
				5	45.7	750	6848	6097.5	1038
1650	5	40	44						
				8	22	1152	3168	2016	3054
1810	11	0	0						
Totals						57732	60786	------>	3054 (Check)

Note: Excavation quantities are based upon cross sections as detailed.
Comment: Balance between cut and fill could be improved but is considered adequate for this preliminary project.

MASS DIAGRAM

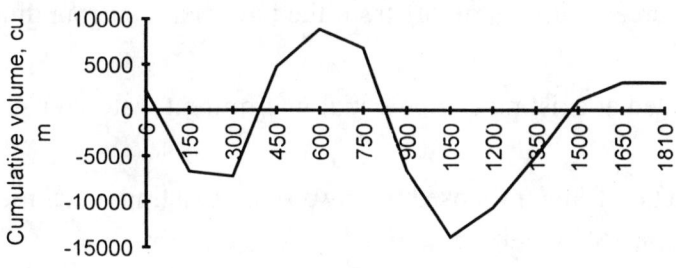

Distance, m

DRAWING NO:	Figure 4-19
TITLE:	EARTHWORKS QUANTITIES AND MASS HAUL
SCALE:	N/A
DESIGNER:	

OUTLINE OF DRAINAGE REQUIREMENTS

The design of the drainage system for any segment of highway is an essential part of the total design process. The complete drainage design for this project is beyond the scope of the preliminary geometric design presented here. Nevertheless, the location of natural waterways, ditches, culverts, and other major drainage appurtenances, together with streams and channels, should be shown. This will ensure that the geometric design and selected route is practicable with regard to drainage requirements and that the highway can be efficiently drained without adversely affecting the natural drainage patterns of adjacent areas, and without requiring costly design features.

Elements of the Drainage System -- For this project, the major features are shown in Figure 4-20 and are briefly described below.

1. Catchment areas and areas of runoff as they relate to the location of the highway.

2. Location of natural streams and channels as well as active streams and the location of dry stream beds and other potentially intermittent watercourses.

3. Locations of the main elements of a potential drainage system including:

   a) ditches to intercept surface runoff from the pavement, surrounding terrain, and side slopes in cuts
   b) culverts located at existing and potential intermittent watercourses.

4. Potential impacts of the proposed highway on existing catchment areas and the drainage of existing highways.

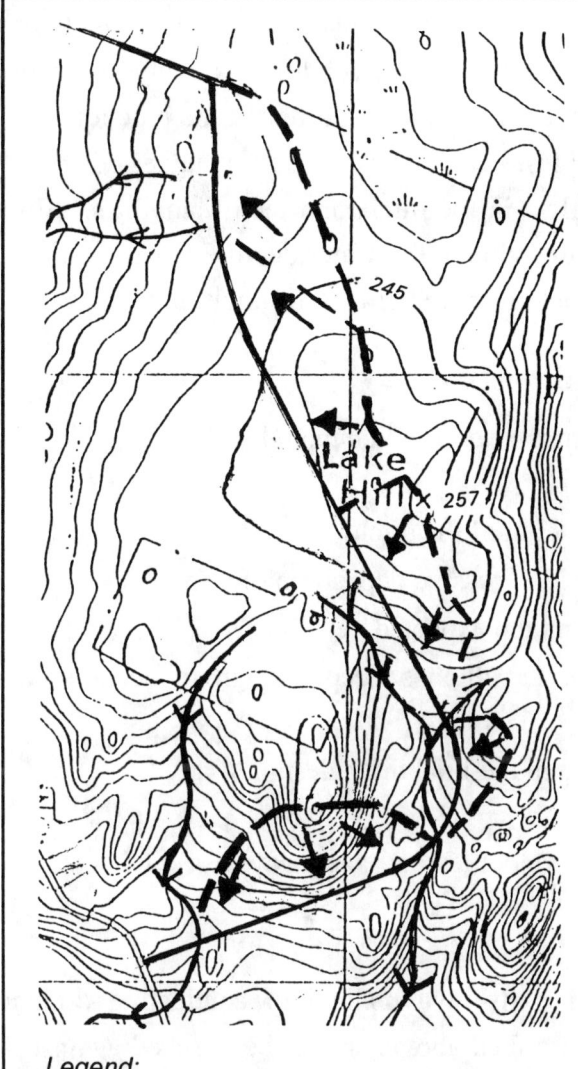

NATURAL DRAINAGE PATTERNS

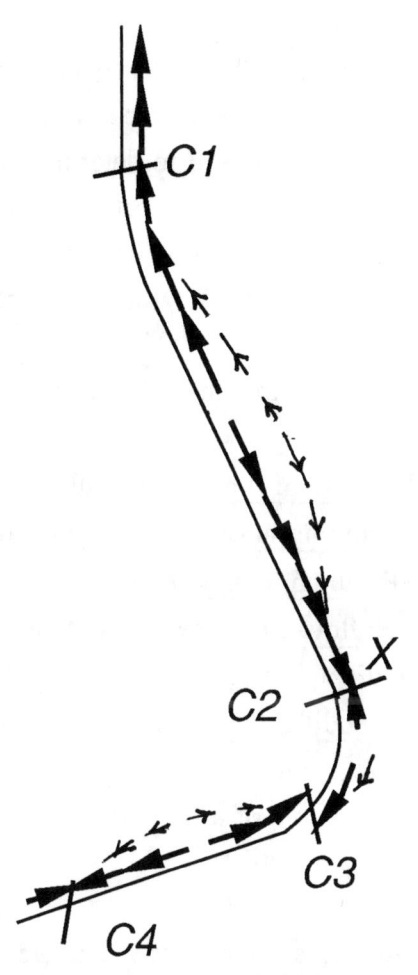

PROPOSED MAJOR DRAINAGE FEATURES

Note: Connections with existing ditches and waterways shall be based upon field inspection and appropriate regulations and design analysis. Interceptor ditches locations based upon profiles and cross sections.

DRAWING NO:	Figure 4-20
TITLE: MAJOR DRAINAGE FEATURES	
SCALE(S):	As shown
DESIGNER:	

Check of Ditch Adequacy -- To check the adequacy of a typical ditch cross section, the location of a ditch where the "worst" case flow condition is likely to occur is identified. This will usually occur for the ditch location -- often at the junction with a culvert or stream -- associated with the largest catchment area. It is suggested that the rational formula be used to determine the flow in the ditch, and then Manning's formula for a specified ditch cross section in order to estimate the water depth, velocity, and type of flow. This can then provide a guide for determining the type of ditch lining to be used and whether problems of silting may occur.

Typical checks on design should determine items which include:

- Is the depth of flow less than the usable depth of the ditch?
- Is the minimum velocity adequate to avoid excessive deposition of sediment?
- Are the ditch linings adequate to withstand the expected velocities?
- Is the flow rapid or tranquil, and are the ditches deep enough to accommodate hydraulic jumps?
- What modifications should be considered in the final ditch design, if any?

An example of a partial check on the velocities and adequacy of a ditch system is shown in Figures 4-21 and 4-22.

Usually, each highway jurisdiction has its own method of conducting drainage design. The proposed preliminary layout, described above, should be checked against the procedures of the authority having jurisdiction, or other acceptable practices, and any modifications should be made where necessary.

Culverts -- The location of culverts, predominantly where stream or intermittent water channels pass under the highway, is shown. Culverts are not dimensioned in this project. An assumed size of 2 m diameter concrete or corrugated steel culverts, each 50 m long, is used for cost-estimating purposes but dimensions in practice should reflect the size of the existing watercourse, plus likely additional flow due to the highway and the cross section configuration.

DITCH CROSS SECTION DESIGN CHECK

Approach

The ditch cross sections with Manning's formula characteristics shown on the next page
are to be checked for adequacy at locations that will be most likely to experience large
volumes of runoff or steep slopes, so that recommendations may be made for design
changes, if any.

Identification of catchment areas and ditch locations for design checks:
Several ditch locations that may experience high volumes and/or velocities are at the
culverts shown in the diagram below. Point X in particular will drain a large catchment
area and also has the portion of the route with the steepest (approximately 5.7%) ditch
slope that will connect with culvert C2.
In order to illustrate the process for checking the adequacy of a ditch, we will present
the calculations and conclusions for point X only.

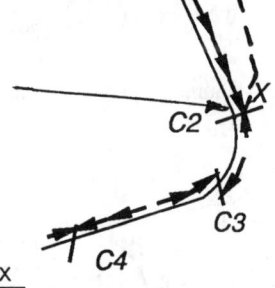

Largest catchment area and
ditch slope, based upon
inspection of previous drawing,
profile, and cross sections

Point X: ditch connects
with culvert

At Point X

Estimation of flow:
Rational formula, Q =0.0028CIA, where Q is in cu m/sec, I is in mm/sec, and A is in hectares

C, forest.................	0.2	from table of runoff coefficients (assume ave. value for forest).
C, asphalt	1	ditto, for asphalt surfaces of pavement and shoulder.
I........	73.15	mm/hr (10 yr, 30 min duration storm, central Massachusetts area. as determined from attached FHWA rainfall chart).
A, forest.........	7.69	Hectares, as measured from the map showing the catchment area.
A, asphalt	0.35	Hectares, half of pavement width plus one shoulder width x approx. 490 m.

Q = 0.0028CIA =	((0.0028 x 0.2 x 7.69) + (0.0028 x 1 x 0.35)) x 73.15....... =	0.39 cu m/sec

Check of ditch adequacy (Manning's formula chart, next page):
Ditch slope (assume approx. equal to the grade near culvert C3) = 5.7%   (0.057m/m)
From chart, Velocity is approximately 1.8 m/sec, Depth is 140 mm and flow is rapid.

Conclusions:
Due to the rapid flow, the ditch must be redesigned; the lining should be rip-rap
or other abrasion resistant lining specified; the slope reduced; or other
individual or combinations of actions taken to provide a satisfactory
ditch in the final design.

DRAWING NO:	*Figure 4-21*
TITLE: *DITCH DESIGN CHECK*	
SCALE(S):	*N/A*
DESIGNER:	

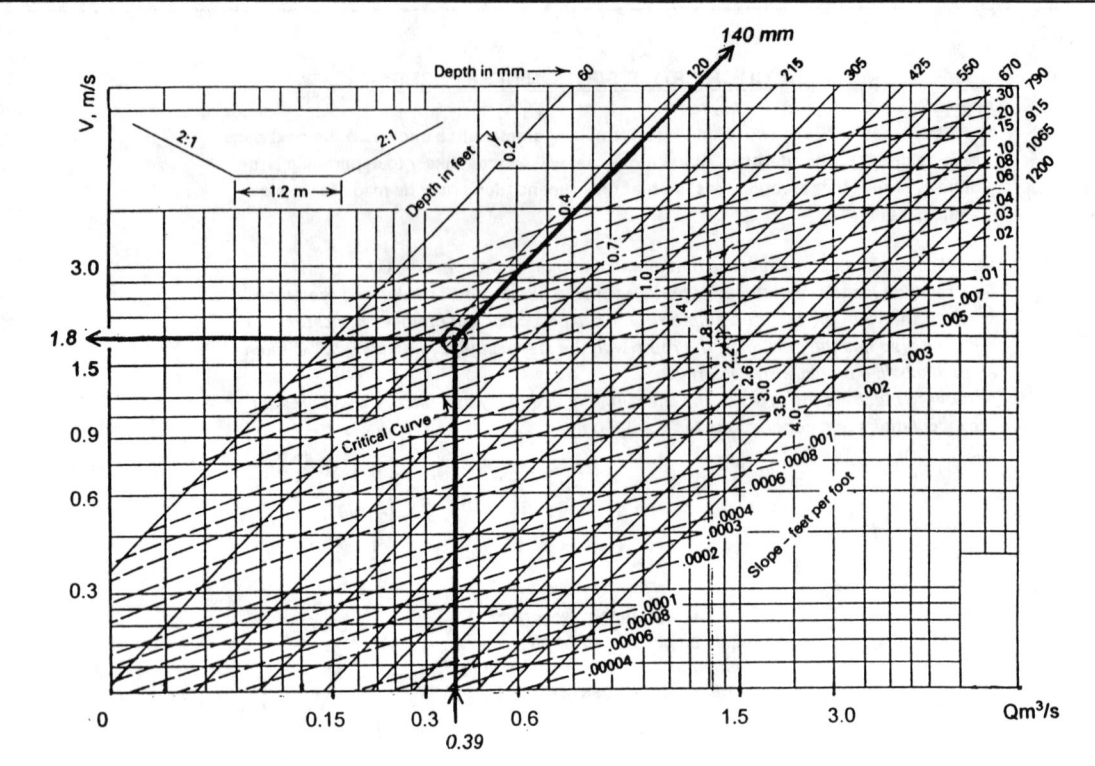

Graphical solution for Manning's equation for channel section shown. Source: Federal Highway Administration.

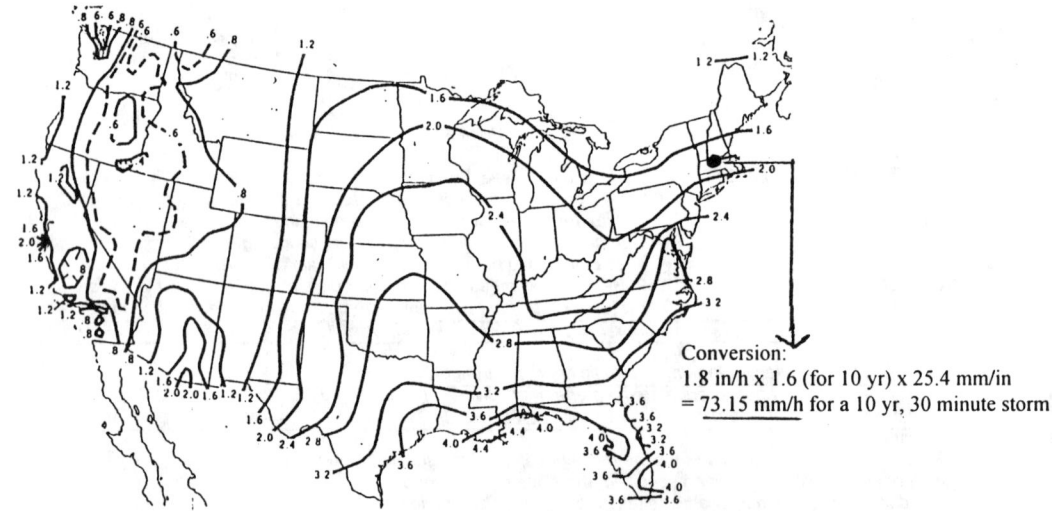

Conversion:
1.8 in/h x 1.6 (for 10 yr) x 25.4 mm/in
= 73.15 mm/h for a 10 yr, 30 minute storm

Map of the contiguous United States, showing 2-year, 30 minute duration rainfall intensity, in/hr. Multiplicative factors for other intensities: 1.44 (15-min); 0.60 (60-min); 0.4 (120 min). Multiplicative factors for recurrence: 0.75 (1-year); 1.6 (10-years) 1.9 (25-years); 22.2 (50-years). Source: Federal Highway Administration.

DRAWING NO:	FIGURE 4-22
TITLE:	
	DITCH DESIGN DATA
SCALE(S):	N/A
DESIGNER:	

INTERSECTION DESIGN

The curve radii and major pavement markings should be shown on the drawing This is not shown here, and is left as an exercise. It is suggested that the radii suitable for a WB-15 design vehicle be used. Appropriate curve radii and related dimensions are shown I Chapter 2. The items to be shown on the intersection design plans are:

- centerline of existing and proposed highways and end points of proposed highway.

- edges of pavements, shoulders, and related features, with appropriate dimensions.

- Compass bearings of existing and proposed highways.

- pavement markings and signs.

We are assuming that in the rural location used in this example, and with the relatively low traffic volumes anticipated, a "T" intersection with a stop sign is warranted. No additional lanes are required at the intersection.

CONSTRUCTION COST ESTIMATE

The preliminary cost estimate provides the engineer, administrators, and others interested in the development of the highway system with an indication of the likely capital cost of the facility. At this stage, the cost estimate is approximate, the total number of items being limited to 12. Each of these items, however, includes many sub-items which would be listed in detail in the final design.

The cost estimate provided in Figure 4-23 indicates the major items, with the dimensions and unit price of each. The unit prices include the materials cost, the contractor's labor, overhead and profit, and a 10% contingency. The unit costs shown are illustrative and based upon those detailed in a number of commercially available construction price publications and those of various states. In practice, typical unit costs would be developed by the agency undertaking the design, based upon local sources of information, and would be adjusted for the year of application by means of construction cost indicators. Also note that this estimate should not be considered a bid price, but an

PRELIMINARY CONSTRUCTION COST ESTIMATE

ITEM	DESCRIPTION	DIMENSIONS (All dimensions in meters)			QUANTIT	UNITS	$ COST/ UNIT (10)	TOTAL COST ($)
1	Clear and grub trees, etc. (1)	Length 1,800	Ave. Width 41	Depth -	7	Hectares	12802.33	94,120
2	Strip and stockpile topsoil (1)	Length 1800	Ave. Width 41	Depth 0.6096	44,816 cu m	cu m	3.47	155,328
3	Excavate in rock and remove (2)	Total Exc'n 54000	% Rock 30		16,200	cu m	32.70	529,689
4	Cut, fill and compact soil (Use largest of soil exc'n or fill for cost est.)	Total Exc'n Rock Exc'n Soil exc'n or Fill			54,000 cu m 16,200 cu m 37,800 From earthw		10.93	412,968
5	Pavement - Gravel Sub-base (3)	Length 1800	Width 15.24	Depth 0.3048	8,361	cu m	10.38	86,829
6	Pavement - Base and surface (4)	Length 1800	Width 12.192	Depth -	21,946	sq m	15.01	329,407
7	Topsoil and seed Incl. ditches (5)	Length 1800	Width 24.384	Depth -	43,891	sq m	0.60	26,248
8	Interceptor ditches (6)	Length 762	Width -	Depth -	762	m	18.04	13,750
9	Culverts (7)	Length 30.48	Number 4	Depth -	122	m	882.55	107,600
10	Guard rail (8)	Length 200	Number 10	Depth -	2,000	m	61.02	122,047
11	Pavement markings (9)	Length 1800	Number 4	Depth -	7,200	m	0.95	6,850
12	Boundary fence on right-of-way	Length 1800	Number 2	Depth -	3,600	m	32.15	115,748

Total dollar cost:	2,000,584
$ Cost per kilometer of highway:	1,111,436

Note: Land costs are not included
(1) Average width is based upon average of cross section widths plus 3 m each side of right-of –way lines.
(2) Total excavation is highest of cut or fill volumes from Figure 4-19.
(3) Assumes that the base extends the full width between the fore slope on each side of the road.
(4) Assumes coverage of pavement and shoulders.
(5) Width is based upon average of ditch and slope widths shown on cross sections.
(6) Based upon approximate lengths of cut slopes intercepting runoff areas.
(7) Based upon approximate lengths estimated from profile and cross sections.
(8) Based upon examination of embankments shown on cross sections.
(9) Assumes two edge stripes and double centerline plus 100 m of miscellaneous markings.
(10) Unit costs are illustrative only, and should be updated to reflect current prices and local conditions.

DRAWING NO:	FIGURE 4-23
TITLE:	PRELIMINARY COST ESTIMATE
SCALE(S):	N/A
DESIGNER:	

engineer's preliminary estimate. Upon issuance of the final design for bids, contractors would provide their own unit prices and total estimates of the construction cost in accordance with their own procedures and marketing strategies. Land costs are not included in the estimate because these vary widely and can be included when final designs are completed. No utilities such as water, power, or other lines are included in this project but could, in practice, be important.

As can be seen from the cost estimate, the total estimated cost of the project is approximately $2 million, or over $1 million per km. Examination of the major cost items indicates that excavation in rock constitutes a considerable portion of this total. Given the obviously steep terrain and other characteristics of the topography, it is reasonable to assume that a considerable amount of rock excavation would be necessary in this case.

SUMMARY OF LIKELY ENVIRONMENTAL ISSUES

A list of environmental issues should be prepared at the preliminary design stage, as shown in Figure 4-24. It is suggested that the major items be compiled, accompanied by applicable notes on each, as soon as the proposed route is reviewed. This will form the basis for a more detailed environmental assessment to be conducted as the project is developed. A list of items likely to be of interest at this preliminary stage, and which should be enlarged as the project proceeds, is as follows:

1. Probable impact of the facility on the environment:
   - Natural environment, such as ecological and visual impacts
   - Relocation or disruption of human activities
   - Recreation for local and other populations
   - Air quality impacts
   - Noise impacts
   - Water quality impacts
   - Construction impacts

2. Probable unavoidable adverse environmental impacts

3. Impacts on cultural and historic sites and attributes of the area

AREA OF ENVIRONMENTAL CONCERN	PROBABLE IMPACTS ASSOCIATED WITH PROPOSED ROUTE	POTENTIAL MITIGATING MEASURES
A. Continuous Impacts		
Wildlife	*Loss of habitat and interrupted movement patterns. Noise and increased human activity may impact adjacent wildlife populations*	*Route modifications resulting from detailed wildlife habitat analysis*
Plants, vegetation	*Loss and/or reduction of species*	*Route modification, possible design changes*
Wetlands	*Highway drainage runoff in streams may degrade wetland quality*	*Provide catchbasins and/or settling ponds to reduce silts and other pollutants*
Air pollution	*Vehicle emissions will add to areawide air pollutant load and also affect receptors in vicinity. Rural area should assist in dispersion of pollution*	*Encourage use of high occupancy vehicles for visitors*
Noise pollution	*May adversely impact human quality of life and wildlife activities*	*Construct barriers to limit adverse impacts where possible. Restrict hours of opening of state reservation*
Drainage patterns	*Changes in runoff patterns may add or decrease amounts of water to sensitive areas*	*Attempt to maintain drainage patterns adjacent to and near the proposed route*
Energy use	*Increases or decreases in energy use may result from changes in travel patterns resulting from the new road*	*Resulting from an areawide analysis of the travel patterns, mitigating measures such as car and van pooling and use of buses may help to reduce overall energy use*
Esthetics	*Loss of trees in cuts may cause visual "scars" inappropriate to the rural nature of the area*	*Seed and plant shrubs and trees on embankments and cut slopes*
B. Impacts During Construction		
Soil runoff and associated stream and wetland pollution	*Excavation and fill areas with exposed and cutback slopes on hillsides are subject to extensive erosion*	*Undertake temporary soil protection measures and construct temporary drainage channels and flow baffles to reduce runoff velocity*

Note: The above list is intended only to highlight the anticipated major areas of environmental concern, and a more detailed evaluation should be made as the project design is refined.

DRAWING NO:	*Figure 4-24*
TITLE:	
SUMMARY OF POTENTIAL ENVIRONMENTAL CONCERNS	
SCALE(S):	*N/A*
DESIGNER:	

Several texts and other documents related to environmental design were listed in Chapter 1, and these may be consulted as a basis for identifying and addressing potential remedial measures associated with environmental impacts. As stated earlier, the intent here is not to develop a detailed environmental assessment but to draw attention to some of the major features to be addressed in such an evaluation.

One very obvious result of the alignment shown in this exercise is that in some cases the cut and fill segments result in slopes of up to 10 m high. To mitigate the adverse visual and ecological effects of these potential "scars" on the landscape, the planting of species similar to those in the surrounding landscape should be given strong consideration. Another solution would be to ensure that the top edges of the back slopes are rounded to minimize erosion and assist in vegetation growth. The possibility of adjusting the alignment as the design is refined should also be considered.

ECONOMIC COST OF PROJECT

The cost estimate conducted here is based upon the cost estimation method incorporated within the economic analysis methods mentioned in Chapter 2. In order to provide some practice in using the AASHTO method of analysis, and enable a valid economic comparison to be made between student projects, the approach is based upon present worth analysis of construction (investment) cost, maintenance cost, and car users' operating costs only. Simplifying assumptions include the following:

- The costs of accidents is likely to be similar for each alternative due to each having somewhat geometrics and only one intersection with the same road. Accident costs, therefore, have not been included as an item that would show significant differences between alternatives in this case.

- The cost of users' travel time is unlikely to differ significantly between alternatives, and the value of travel time for essentially leisure pursuits is likely to be low, and has not been included.

- The operating cost of trucks and commercial vehicles has not been included because of the expected low percentage in the traffic stream and the minimal difference in costs between each alternative.

- The salvage value of each project is likely to show little difference, and is not included for comparison purposes.

The economic cost computations are summarized in Figure 4-25 where they are assembled in a format that will assist the use of either manual or computerized worksheets. Figure 4-26 shows the use of the ASSHTO passenger vehicle operating costs and the values that apply in this example for inputs to the calculations. The main features of Figure 4-25 are:

- A graphical, keyed summary of the main geometric features (horizontal and vertical alignment) and the necessary input parameters.

- Vertical and horizontal alignment summaries corresponding to the graphical summary, above, and with values derived from the alignment calculations conducted earlier in the project. Note that the total distance in the vertical alignment must equal the total length of the route.

- Construction cost, the total amount of which is assumed to occur in year 1, and therefore is not modified by a PW factor.

- Maintenance cost, assumed here to be 10% of construction cost throughout the 20-year life of the project, and therefore converted to a present worth by application of the series present worth (SPW) factor.

- Annual user (vehicle operating costs) based upon Figure 4-26.

- Present worth of vehicle operating costs, estimated by applying the SPW factor to the annual vehicle operating costs.

- Summary and total of all PWs, i.e., for construction, maintenance, and vehicle operating costs.

ECONOMIC COST ANALYSIS

INTRODUCTION
Economic analysis method - present worth of construction, maintenance, operation, interest and user costs.

Source: "A Manual on User Benefit Analysis of highway and Bus-Transit Improvements" (AASHTO, 1997) - Chapter 2, Ref. 7.

Design features for economic analysis
SUMMARY OF MAIN FEATURES

Horizontal Alignment	Vertical Alignment	Input Parameters	
		ADT: Estimated 2000 in yr. 2000, and 3500 in yr. 2020	
		Average ADT =(2000 + 3500) / 2 =	2750
		Update factor per Consumer Price Index (CPI)	2.3
		Days per year for recreational facilities	120
		(weekends, holidays and vacation periods)	

VERTICAL ALIGNMENT SUMMARY

Curve	Va	Vb	Vc	Vd	Ve	Vf	Vg	Vh	All	Units	Source
Grade	2	2.8	5.7	-2.3	-0.7					%	Vertical alignment calculations
Length	122	396	579	579	140				1816	m	Vertical alignment calculations
											(Total equals length of project)

HORIZONTAL ALIGNMENT SUMMARY

Curve reference	Ha	Hb	Hc	Hd	He	Hf	Hg	Hh	All		
Radius	157	514								m	Horizontal alignment calculations
Degree of curve	11.2	3.4								degrees	Horizontal alignment calculations
Length	269	233							502	m	Horizontal alignment calculations

[1]	CONSTRUCTION				
[1.1]	Construction cost	2000584	$	From construction cost estimate	
[2]	MAINTENANCE				
[2.1]	Annual maintenance	200058.4	$/year	Assumed 10% of constr'n cost	
[2.2]	SPW factor	13.59	constant	4% for 20 years. SPW table in Appendix	
[2.3]	PW of maintenance	2718794	$	[2.1] x [2.2]	

[3] VEHICLE OPERATING COSTS

[3.] Vertical alignment (grade) costs. Begin point A, toward point 1.

		Va	Vb	Vc	Vd	Ve	Vf	Vg	Vh	All	Units	Source
[3.1.1]	Grade ref.											
[3.1.2]	Grade	2	2.8	5.7	-2.3	-0.7					%	Above
[3.1.3]	Grade unit operating cost	49	51	59	35	41					$/1,000 veh-km	From Figure 4-26
[3.1.4]	Grade length	0.12	0.40	0.58	0.58	0.14				1.82	km	Above/1000
[3.1.5]	Grade operating cost	14	46	79	47	13					$	[3.1.3] x [3.1.4] x CPI factor, above
[3.1.6]	One-way ADT	1375	1375	1375	1375	1375					Vehicles	Input parameter, above for 1-way only
[3.1.7]	Days per year	120	120	120	120	120					days	Input parameter, above
[3.1.8]	Ave. one-way traffic in 1 year	165	165	165	165	165					1,000 veh.	[3.1.6] x [3.1.7]/1000
[3.1.9]	Annual veh. operating cost	2269	7664	12964	7691	2178				32766	$/year	Sum for all grades

[3.2] Vertical alignment (grade) costs. Begin at point 1, toward point A.

		Ve	Vd	Vc	Vb	Va	Vf	Vg	Vh	All	Units	Source
[3.2.1]	Grade ref.											
[3.2.2]	Grade	0.7	2.3	-5.7	-2.8	-2					%	Above
[3.2.3]	Grade unit operating cost	47	50	35	35	35					$/1,000 veh-km	From Figure 4-26
[3.2.4]	Grade length	0.14	0.58	0.58	0.4	0.12				1.82	km	Above/1000
[3.2.5]	Grade operating cost	15.134	66.7	46.69	32.2	9.66					$	[3.2.3] x [3.2.4] x CPI factor, above
[3.2.6]	One-way ADT	1375	1375	1375	1375	1375					Vehicles	Input parameter, above for 1-way only
[3.2.7]	Days per year	120	120	120	120	120					days	Input parameter, above
[3.2.8]	Ave. one-way traffic in 1 year	165	165	165	165	165					1,000 veh.	[3.2.6] x [3.2.7]/1000
[3.2.9]	Annual veh. operating cost	2497	11006	7704	5313	1594				28113	$/year	Sum for all grades

[3.3] Horizontal alignment cost

		Ha	Hb	Hc	Hd	He	Hf	Hg	Hh	All	Units	Source
[3.3.1]	Curve reference											
[3.3.2]	Degree of curve	11.2	3.4								Degrees	Above
[3.3.3]	Veh unit operating cost	26	3								$/1,000 veh-km	From Figure 4-26
[3.3.4]	Curve length	0.269	0.233								km	Above/1000
[3.3.5]	Veh operating cost	6.994	0.699								$	[3.3.3] x [3.3.4] x CPI factor, above
[3.3.6]	ADT	2750	2750								Vehicles	Input parameter, above
[3.3.7]	Days/year	120	120								days	Input parameter, above
[3.3.8]	Average traffic per year	330	330								1,000 veh.	[3.3.6] x [3.3.7]/1000
[3.3.9]	Annual veh. operating cost	2308	231							2539	$/year	Sum for all grades

[4]	PRESENT WORTH OF VEHICLE OPERATING COSTS				
[4.1]	Total curve and grade oper. costs	63418	$/year	[3.1.9] + [3.2.9] + [3.3.9]	
[4.2]	SPW factor	13.59	Constant	4% for 20 years - SPW table in Appendix	
[4.3]	PW of vehicle operating costs	861852	$	[4.1] x [4.2]	
[5]	SUMMARY				
[5.1]	PW construction cost	2000584	$	[1.1]	
[5.2]	PW of maintenance costs	2718794	$	[2.3]	
[5.3]	PW of operating costs	861852	$	[4.3]	
[6]	**Present Worth of project ($)**	**5,581,229**	$	[5.1] + [5.2] + [5.3]	

(1) In accordance with CPI cost update procedure in Chapter 2, Ref. 7.

DRAWING NO:	*FIGURE 4-25*
TITLE:	
	ECONOMIC COST ANALYSIS WORKSHEET
SCALES:	*Not applicable*
DESIGNER:	

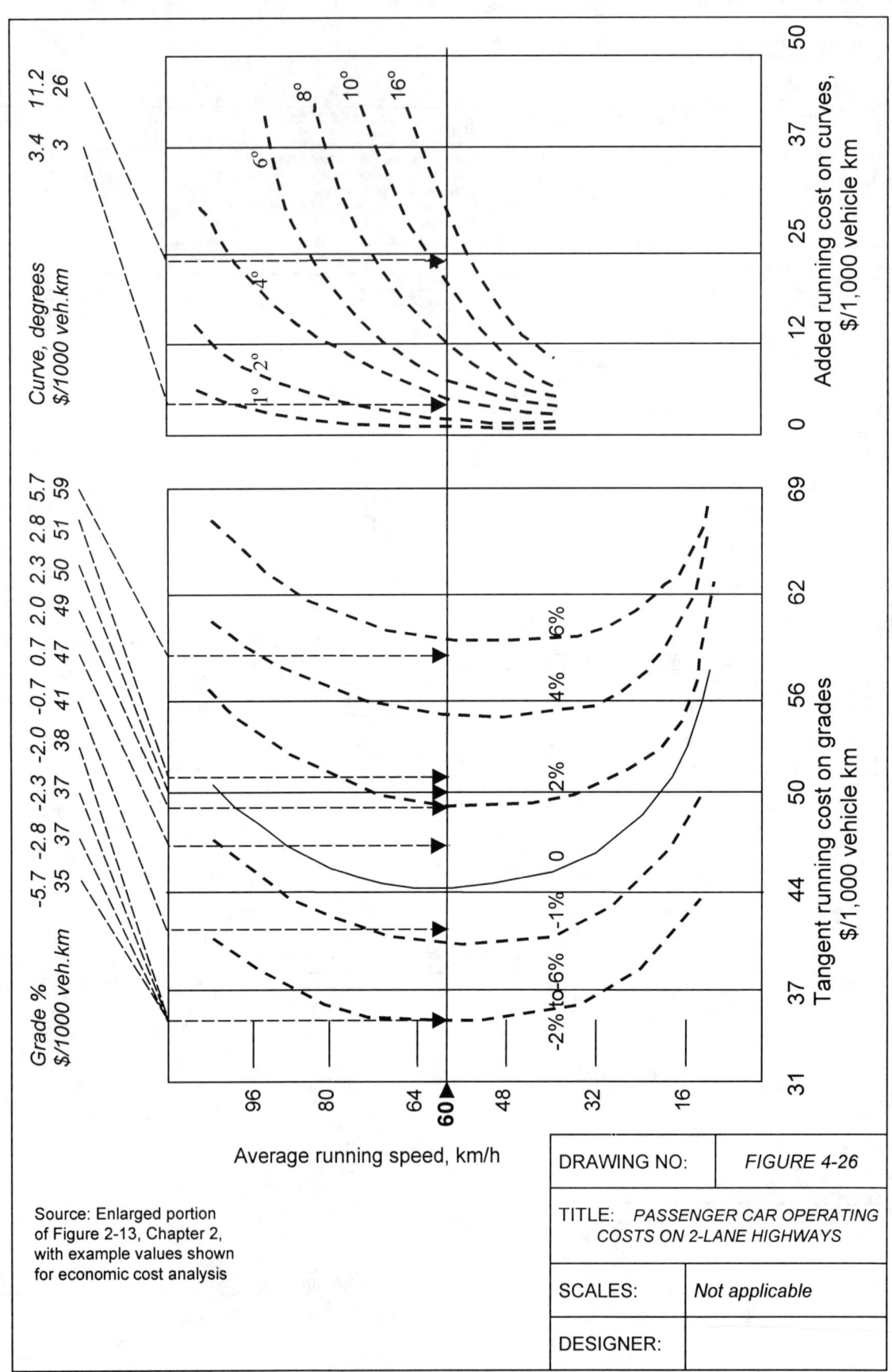

Source: Enlarged portion
of Figure 2-13, Chapter 2,
with example values shown
for economic cost analysis

DRAWING NO:	FIGURE 4-26
TITLE:	PASSENGER CAR OPERATING COSTS ON 2-LANE HIGHWAYS
SCALES:	Not applicable
DESIGNER:	

It can be seen that the total PW amounts to nearly $5.6 million -- almost three times the construction cost alone -- and that the maintenance cost amounts to nearly half the construction cost. Note that because there is no existing highway to provide a comparison with the proposed highway a benefit-cost analysis in the traditional sense cannot be made. The economic cost as calculated, however, provides a cost measure that can be compared with other routes for the same end-points. Thus an indication of the preferred route from an economic cost point of view can be made. The economic results should be arrayed with the other features of the project in order to provide a greater perspective at this stage as well as assisting with comparisons with alternative routes between the same beginning and end points. This is done below.

SUMMARY OF KEY TECHNICAL FEATURES

The summary should consist of a tabular format outlining the specified and proposed parameters, cost estimates, economic and other relevant information. The summary for the project is shown in Figure 4-27. The number of columns may be increased to the right to provide a direct comparison with alternative projects.

TECHNICAL SUMMARY

Project No. *A2/99*   Name of designer   *J.G.S.*   Date *18 Aug.99*

ITEM	SPECIFICATION PARAMETERS	VALUES USED IN, AND RESULTING FROM, ACTUAL DESIGN
Total length	*Not specified*	*1816 m*
Maximum grade, generally	*7%*	*5.70%*
Maximum grade, at intersections	*2%*	*2%*
Maximum cut depth	*7 m*	*7 m*
Maximum fill height	*7 m*	*7 m*
Intersection angles	*90 degrees plus or minus 15 degrees*	*90 degrees at Point A* *86 degrees at Point 1*
Total cut (cu m)	*Not specified*	*57732 cu m*
Total fill (cu m)	*Not specified*	*60786 cu m*
Excess cut or fill	*Not specified*	*3054 cu m*
Capital cost	*Not specified*	*$2,000,584*
Capital cost per kilometer	*Not specified*	*$1,111,436*
Economic cost (present worth)	*Not specified*	*$5.59 million*

DRAWING NO:		*Figure 4-27*
TITLE: *SUMMARY OF TECHNICAL FEATURES*		
SCALE(S):		*N/A*
DESIGNER:		*J.G.S.*

Chapter 5

Preliminary Highway Geometric Design Projects

This chapter specifies the requirements for undertaking several student term projects similar in scope to that outlined in Chapter 4. Additional information to assist in monitoring the progress of the projects is contained in the solutions manual available with this book.

ROUTE SELECTION AND DESIGN PROJECTS

The description, objectives, and specifications of the projects are presented in Table 5-1, and the locations of the project's end points are shown in Figures 5-1 and 5-2. There is a total of 33 separate projects, each of which may be conducted by an individual or group. If desired, of course, other end points, maximum grades, and other controls for the proposed highway may be investigated in addition to those suggested here, or other localities and maps may be used. Note, however, that specification of excessively stringent controls may result in there being no technically feasible route available between the selected end points.

It often advantageous to photographically enlarge the maps in Figures 5-1 and 5-2 to twice their size, or more, to facilitate scaling and general interpretation. The scale should, of course, be adjusted accordingly. It is usually best to photographically enlarge the scale and the map together to ensure that there is no discrepancy between them.

BACKGROUND: Interest has been expressed in a program to establish visitor centers in scenic areas. In order to obtain an indication of potential access highway costs for these centers, the locations of several centers have been identified on maps in order to provide "case studies" as a means of estimating typical costs of a state-wide program. In addition, an approximate indication of the differences in construction, maintenance and user costs resulting from maximum grades of 6%, 7%, and 8%, with corresponding maximum cut and fill heights of 7 m, 8.5 m and 10 m is required. Examples of case studies are the visitor centers and scenic lookouts on the promontories located at Point 1 in Figure 5-1(A) and 5-1(B) and Point 2 in Figure 5-2. Several possible end points for the access highway between the center and the adjacent highway network have been identified for the case studies. The end points are designated A, B, C, D, and E in Figures 5-1(A) and 5-1(B) and A, B, C, D, E, and F in Figure 5-2.

OBJECTIVES: Preliminary highway designs and economic costs of construction, maintenance, and user operations are needed, together with an initial list of environmental concerns. Design and costing of the visitor center are being done separately, and are not included as part of this project.

SCOPE OF PROJECT: Conduct a preliminary route selection and design for each of the possible route end combinations and grades (See Project Ref. Nos., below), for each of the specified grades, for a two-lane highway in accordance with AASHTO "rural collector" highway design standards; estimate the construction, maintenance, and users' operational costs of each; and compare the costs of the different projects. The content, format, and general presentation of the report should follow the guidelines of Chapter 4 of this book, but may contain greater detail if time, resources, and interest permit. A listing of the main items in the required content of the report is shown also in Chapter 4, and summarized in Table 5-2. This table may be used as the basis for the table of contents of your report.

	PROJECT REFERENCE NUMBERS (see note below)					
	Figure 5-1			Figure 5-2		
ITEM	1-A-6	1-A-7	1-A-8	2-A-6	2-A-7	2-A-8
	1-B-6	1-B-7	1-B-8	2-B-6	2-B-7	2-B-8
	1-C-6	1-C-7	1-C-8	2-C-6	2-C-7	2-C-8
	1-D-6	1-D-7	1-D-8	2-D-6	2-D-7	2-D-8
	1-E-6	1-E-7	1-E-8	2-E-6	2-E-7	2-E-8
Design Designations:						
V [km/h]				60		
ADT (Current Year)				0		
ADT (Future Year)	◄	◄	◄	4,000	►	►
K [%]				10		
D [%]				60		
T [%]				2		
Design Controls:						
Max. grade, except at intersections [%]	6	7	8	6	7	8
Max. fill height and cut depth at centerline [m]	10	8.5	7	10	8.5	7
Min. grade (all locations) [%]				0.5		
Max. vertical curve K value, drainage criterion	◄	◄	◄	51	►	►
e (max)				6		
Design vehicle				WB-15		
Other Conditions:						
Similar to "Intersection Geometrics" and "Other Conditions", Figures 4-2 and 4-3, Ch. 4.						

NOTE: 1 - A - 7, etc

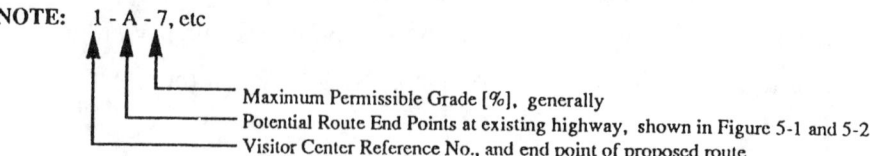

Maximum Permissible Grade [%], generally
Potential Route End Points at existing highway, shown in Figure 5-1 and 5-2
Visitor Center Reference No., and end point of proposed route

TABLE 5-1

PROJECT DESCRIPTION AND PARAMETERS

Location of visitor center, finished
surface elevation 305 m, shown thus:

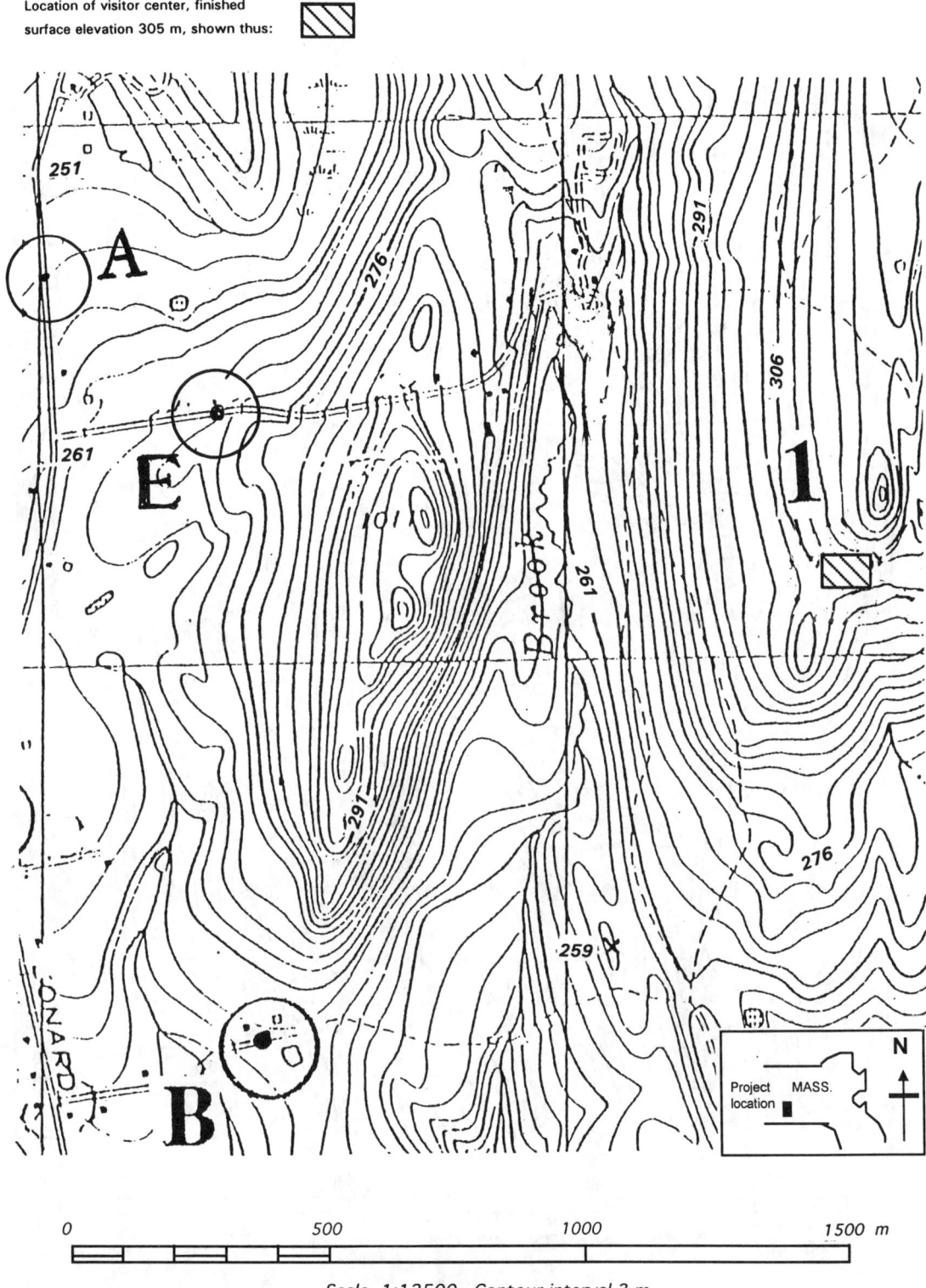

Scale, 1:12500, Contour interval 3 m

FIGURE 5-1(A)
ALTERNATIVE PROJECT ROUTE END POINTS

133

Location of visitor center, finished
surface elevation 305 m, shown thus:

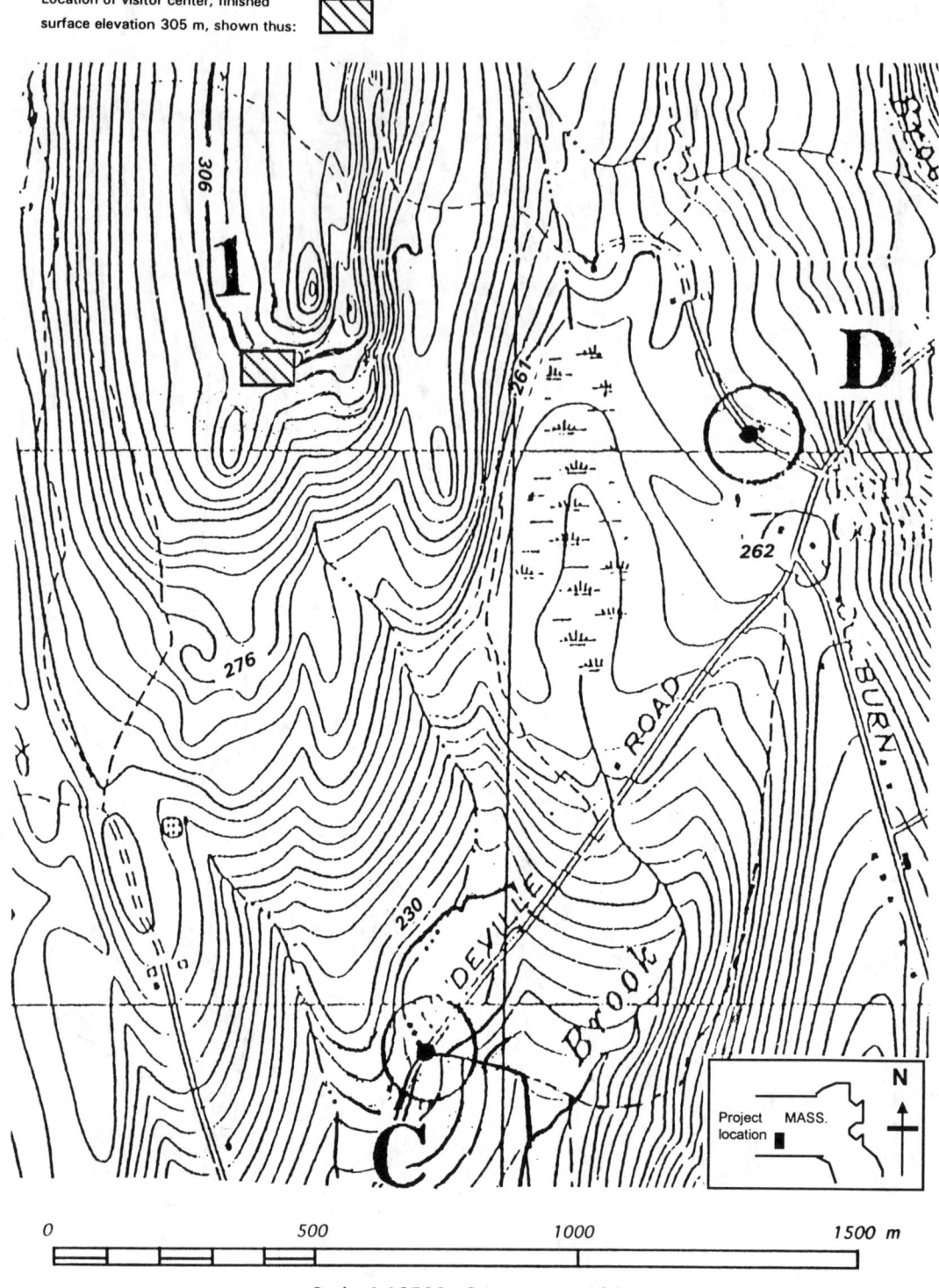

Scale, 1:12500, Contour interval 3 m

FIGURE 5-1(B)
ALTERNATIVE PROJECT ROUTE END POINTS

Location of visitor center, finished surface elevation 322 m, shown thus: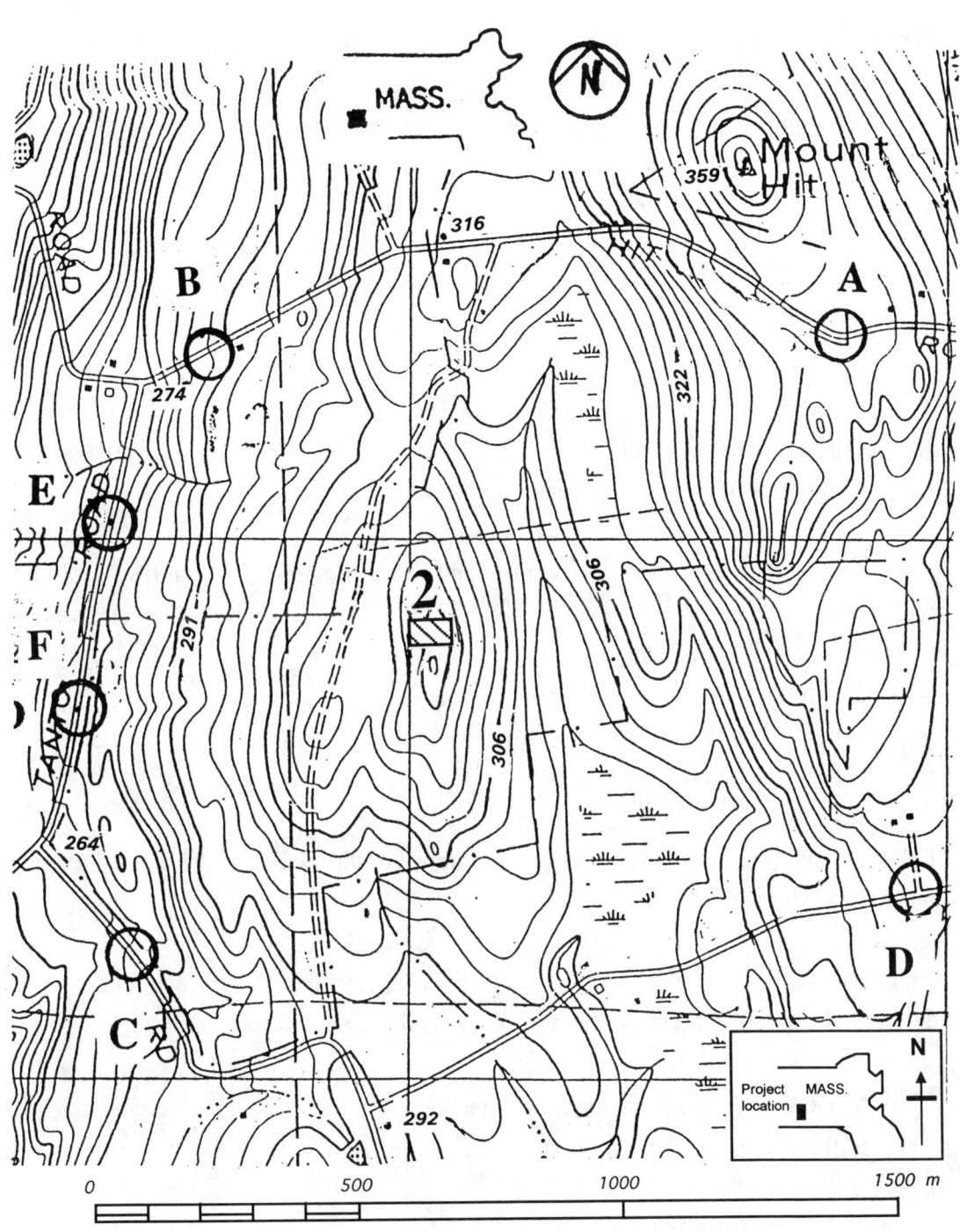

FIGURE 5-2
ALTERNATIVE PROJECT ROUTE END POINTS

Scale, 1:12500, Contour interval 3 m

REPORT FORMAT AND CHECK LIST

Report Format -- In general, most highway design agencies publish guidelines related to the presentation of highway designs both as a means of increasing the efficiency and comprehension of their design documents, and to assist non-technical interest groups and individuals in understanding the major features of the proposed designs. To reflect this approach to presenting desired information in concise but adequate fashion, and as a means of assisting the comparison of the assigned projects, a suggested format for presenting the design projects is described graphically in Figure 5-3. Key points are as follows:

- The diagrams show the major features of the contents of each double letter sized sheet (approximately A3 in European pages sizes) sheet in the design document, and how the contents may be assembled on each page.

- The type and extent of detail to be provided in the documents would be similar to that shown in Chapter 4.

- Also shown are the suggested scales for the drawings. It may be appropriate to modify these if the selected project is significantly longer or shorter than the 1800 m project shown in Chapter 4.

- Different approaches to the sheet layouts may be adopted if necessary to adapt to unusual conditions.

It should be noted that in practice the design sheets would typically show much more detail, and examples are presented in several of the texts referred to earlier.

Check List -- To assist in ensuring that the main points of each element of the design are addressed, a check list to augment the information provided in the earlier chapters is provided in Table 5-2.

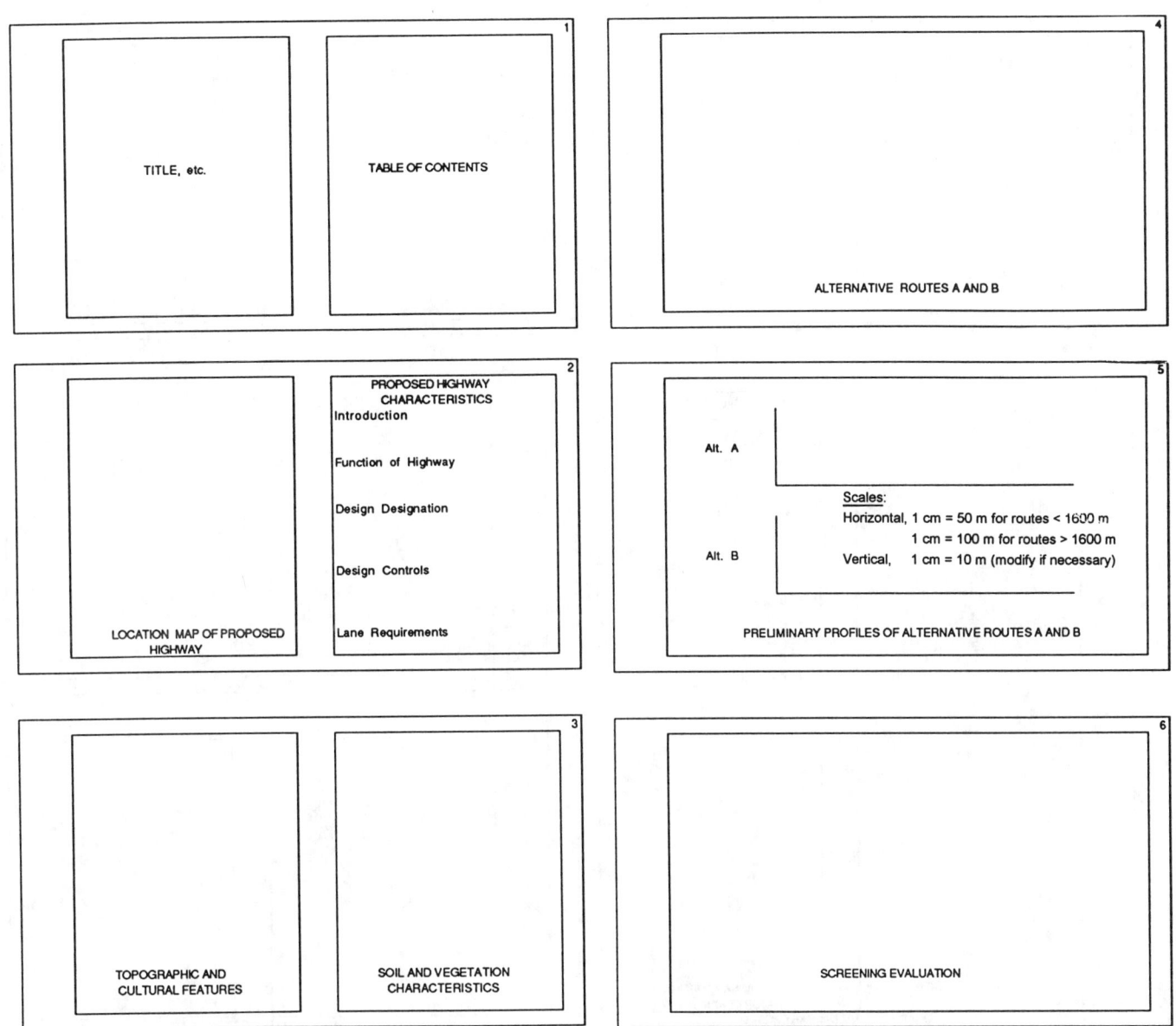

FIGURE 5-3
DESIGN PROJECT REPORT FORMAT
...cont'd

...cont'd
FIGURE 5-3
DESIGN PROJECT REPORT FORMAT
cont'd...

...cont'd
FIGURE 5-3
DESIGN PROJECT REPORT FORMAT
cont'd...

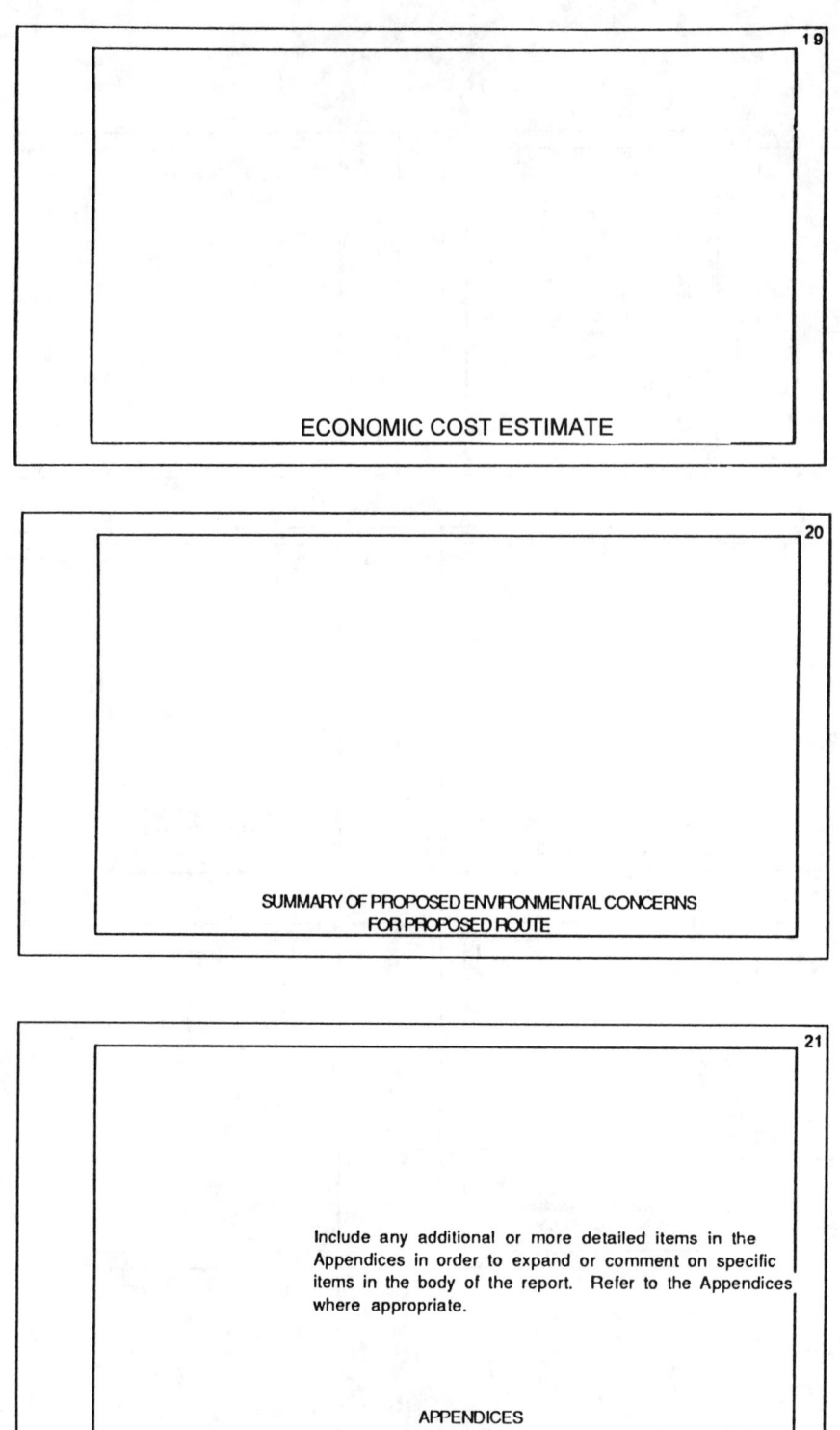

ECONOMIC COST ESTIMATE

SUMMARY OF PROPOSED ENVIRONMENTAL CONCERNS
FOR PROPOSED ROUTE

Include any additional or more detailed items in the Appendices in order to expand or comment on specific items in the body of the report. Refer to the Appendices where appropriate.

APPENDICES

...cont'd
FIGURE 5-3
DESIGN PROJECT REPORT FORMAT

TABLE 5-2
CHECK LIST OF REPORT CONTENTS

- Title Page - project name - author - client - date

- Table of Contents - self explanatory

- Location Map - scale - orientation - adjacent highways - state and regional location

- Function of Proposed Highway - justification for functional classification

- Design Designations and Controls - as listed in Chapters 2 and 4

- Consideration of Alternative Routes -
    - Plan of each potential route - orientation - scale - topographic features - labeled end point locations - selected route - PC and PT locations - curve radii
    - Profiles of each potential route - vertical and horizontal scales - existing ground level - potential highway profile indicating grades and vertical curve lengths for each segment

- Screening of Alternatives - see Chapter 4 example - clearly state the selected route and why selected

- Horizontal Alignment - bearings - PIs - PCs - PTs - deflection angles - total route length - traverse closure - computations and comments

- Profile (Vertical Alignment) of Proposed Highway - see example in Chapter 4 - check total length with horizontal alignment length

- Coordination of Vertical and Horizontal Alignment - see example in Chapter 4 - comments

- Examples of Curve Design - see example in Chapter 4

- Cross Sections - Total of 10 to 12 required for this project - provide station and area for each - show all ditches, including interceptors - check ditch location with drainage plans - check that ground level is correct based upon contours

- Earthworks Quantities and Mass Diagram - see Chapter 4 - comment on findings

- Drainage:
    - Selected Catchment Area - clearly delineate boundaries
    - Plan of ditch, culvert layout - show interceptor ditches also
    - Runoff computation and check of ditch design - see Chapter 4 - comment on findings and potential actions

- Intersection Design - main features as listed in earlier chapters

- Construction Cost Estimate - see Chapter 4

- Economic Analysis - conduct this per example in Chapter 4 - use current or agreed common CPI factor - compare with other route alternatives for identical beginning and end points if available

- Summary of Likely Environmental Issues - address this particular project

- Summary of Key Technical Features - list

Appendices

National Mapping Program

Topographic Map Symbols

National Large Scale Series

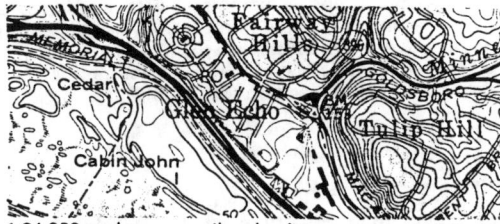

1:24,000 scale—conventional units

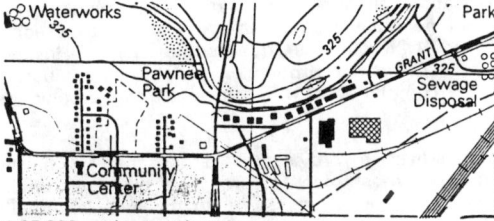

1:25,000 scale—metric units

Provisional edition

U. S. Department of the Interior
Geological Survey
National Mapping Division

Map series and quadrangles

Each map in a U. S. Geological Survey series conforms to established specifications for size, scale, content, and symbolization. Except for maps which are formatted on a County or State basis, USGS quadrangle series maps cover areas bounded by parallels of latitude and meridians of longitude.

Map scale

Map scale is the relationship between distance on a map and the corresponding distance on the ground. Scale is expressed as a ratio, such as 1:25,000, and shown graphically by bar scales marked in feet and miles or in meters and kilometers.

Standard edition maps

Standard edition topographic maps are produced at 1:20,000 scale (Puerto Rico) and 1:24,000 or 1:25,000 scale (conterminous United States and Hawaii) in either 7.5 x 7.5- or 7.5 x 15-minute format. In Alaska, standard edition maps are available at 1:63,360 scale in 7.5 x 20 to 36-minute quadrangles. Generally, distances and elevations on 1:24,000-scale maps are given in conventional units: miles and feet, and on 1:25,000-scale maps in metric units: kilometers and meters.

The shape of the Earth's surface, portrayed by contours, is the distinctive characteristic of topographic maps. Contours are imaginary lines which follow the land surface or the ocean bottom at a constant elevation above or below sea level. The contour interval is the elevation difference between adjacent contour lines. The contour interval is chosen on the basis of the map scale and on the local relief. A small contour interval is used for flat areas; larger intervals are used for mountainous terrain. In very flat areas, the contour interval may not show sufficient surface detail and supplementary contours at less than the regular interval are used.

The use of color helps to distinguish kinds of features:

Black—cultural features such as roads and buildings.
Blue—hydrographic features such as lakes and rivers.
Brown—hypsographic features shown by contour lines.
Green—woodland cover, scrub, orchards, and vineyards.
Red—important roads and public land survey system.
Purple—features added from aerial photographs during map revision. The changes are not field checked.

Series	Scale	1 inch represents approximately	1 centimeter represents	Size (latitude x longitude)	Area (square miles)
Puerto Rico 7.5-minute	1:20,000	1,667 feet	200 meters	7.5 x 7.5 min.	71
7.5-minute	1:24,000	2,000 feet (exact)	240 meters	7.5 x 7.5 min.	49 to 70
7.5-minute	1:25,000	2,083 feet	250 meters	7.5 x 7.5 min.	49 to 70
7.5 x 15-minute	1:25,000	2,083 feet	250 meters	7.5 x 15 min.	98 to 140
USGS/DMA 15-minute	1:50,000	4,166 feet	500 meters	15 x 15 min.	197 to 282
15-minute	1:62,500	1 mile	625 meters	15 x 15 min.	197 to 282
Alaska 1:63,360	1:63,360	1 mile (exact)	633.6 meters	15 x 20 to 36 min.	207 to 281
County 1:50,000	1:50,000	4,166 feet	500 meters	County area	Varies
County 1:100,000	1:100,000	1.6 miles	1 kilometer	County area	Varies
30 x 60-minute	1:100,000	1.6 miles	1 kilometer	30 x 60 min.	1,568 to 2,240
U. S. 1:250,000	1:250,000	4 miles	2.5 kilometers	1° x 2° or 3°	4,580 to 8,669
State maps	1:500,000	8 miles	5 kilometers	State area	Varies
U. S. 1:1,000,000	1:1,000,000	16 miles	10 kilometers	4° x 6°	73,734 to 102,759
U. S. Sectional	1:2,000,000	32 miles	20 kilometers	State groups	Varies
Antarctica 1:250,000	1:250,000	4 miles	2.5 kilometers	1° x 3° to 15°	4,089 to 8,336
Antarctica 1:500,000	1:500,000	8 miles	5 kilometers	2° x 7.5°	28,174 to 30,462

Map series and quadrangles

Each map in a U. S. Geological Survey series conforms to established specifications for size, scale, content, and symbolization. Except for maps which are formatted on a County or State basis, USGS quadrangle series maps cover areas bounded by parallels of latitude and meridians of longitude.

Map scale

Map scale is the relationship between distance on a map and the corresponding distance on the ground. Scale is expressed as a ratio, such as 1:25,000, and shown graphically by bar scales marked in feet and miles or in meters and kilometers.

Standard edition maps

Standard edition topographic maps are produced at 1:20,000 scale (Puerto Rico) and 1:24,000 or 1:25,000 scale (conterminous United States and Hawaii) in either 7.5 x 7.5- or 7.5 x 15-minute format. In Alaska, standard edition maps are available at 1:63,360 scale in 7.5 x 20 to 36-minute quadrangles. Generally, distances and elevations on 1:24,000-scale maps are given in conventional units: miles and feet, and on 1:25,000-scale maps in metric units: kilometers and meters.

The shape of the Earth's surface, portrayed by contours, is the distinctive characteristic of topographic maps. Contours are imaginary lines which follow the land surface or the ocean bottom at a constant elevation above or below sea level. The contour interval is the elevation difference between adjacent contour lines. The contour interval is chosen on the basis of the map scale and on the local relief. A small contour interval is used for flat areas; larger intervals are used for mountainous terrain. In very flat areas, the contour interval may not show sufficient surface detail and supplementary contours at less than the regular interval are used.

The use of color helps to distinguish kinds of features:

Black — cultural features such as roads and buildings.
Blue — hydrographic features such as lakes and rivers.
Brown — hypsographic features shown by contour lines.
Green — woodland cover, scrub, orchards, and vineyards.
Red — important roads and public land survey system.
Purple — features added from aerial photographs during map revision. The changes are not field checked.

Some quadrangles are mapped by a combination of orthophotographic images and map symbols. Orthophotographs are derived from aerial photographs by removing image displacements due to camera tilt and terrain relief variations. An orthophotoquad is a standard quadrangle format map on which an orthophotograph is combined with a grid, a few place names, and highway route numbers. An orthophotomap is a standard quadrangle format map on which a color enhanced orthophotograph is combined with the normal cartographic detail of a standard edition topographic map.

Provisional edition maps

Provisional edition maps are produced at 1:24,000 or 1:25,000 scale (1:63,360 for Alaskan 15-minute maps) in conventional or metric units and in either a 7.5 x 7.5- or 7.5 x 15-minute format. Map content generally is the same as for standard edition 1:24,000- or 1:25,000-scale quadrangle maps. However, modified symbolism and production procedures are used to speed up the completion of U.S. large-scale topographic map coverage.

The maps reflect a provisional rather than a finished appearance. For most map features and type, the original manuscripts which are prepared when the map is compiled from aerial photographs, including hand lettering, serve as the final copy for printing. Typeset lettering is applied only for features which are designated by an approved name. The number of names and descriptive labels shown on provisional maps is less than that shown on standard edition maps. For example, church, school, road, and railroad names are omitted.

Provisional edition maps are sold and distributed under the same procedures that apply to standard edition maps. At some future time, provisional maps will be updated and reissued as standard edition topographic maps.

National Mapping Program indexes

Indexes for each State, Puerto Rico, the U. S. Virgin Islands, Guam, American Samoa, and Antarctica are available. Separate indexes are available for 1:100,000-scale quadrangle and county maps; USGS/Defense Mapping Agency 15-minute (1:50,000-scale) maps; U. S. small scale maps (1:250,000, 1:1,000,000, 1:2,000,000 scale; State base maps; and U. S. maps); land use/land cover products; and digital cartographic products.

Provisional edition maps - metric or conventional units

Metric unit maps

Conventional unit maps

CONTROL DATA AND MONUMENTS

	Conventional	Metric	Provisional
Aerial photograph roll and frame number	Not Shown	Not Shown	3-20
Horizontal control:			
Third order or better, permanent mark	Neace △	Neace △	Neace ⊕
With third order or better elevation	BM △ 148	BM △ 45.1	⊕ Pike BM 45.1
Checked spot elevation	△ 64	△ 19.5	Not Shown
Coincident with section corner	△ Cactus	△ Cactus	⊕ Cactus
Unmonumented	Not Shown	Not Shown	+
Vertical control:			
Third order or better, with tablet	BM × 53	BM × 16.3	BM × 53.4
Third order or better, recoverable mark	× 394	× 120.0	× 393.6
Bench mark at found section corner	BM + 61	BM + 18.6	BM + 60.9
Spot elevation	× 17	× 5.3	× 17
Boundary monument:			
With tablet	BM □ 71	BM □ 21.6	BM ⊞ 71
Without tablet	□ 562	□ 171.3	□ 562
With number and elevation	67 □ 988	67 □ 301.1	67 □ 988
U.S. mineral or location monument	▲	▲	USMM ▲

BOUNDARIES

National	— —— —		
State or territorial	— — —		
County or equivalent	— — —		
Civil township or equivalent	— — —		
Incorporated·city or equivalent	— — —		
Park, reservation, or monument	— — —		
Small park	———————		

LAND SURVEY SYSTEMS

U.S. Public Land Survey System:

Township or range line			
Location doubtful			
Section line			
Location doubtful			
Found section corner; found closing corner	+ | +	+ | +	+ | +
Witness corner; meander corner	WC + MC	WC + MC	WC + MC

Other land surveys:

Township or range line			
Section line			
Land grant or mining claim; monument	— □	— □	— □
Fence line			

ROADS AND RELATED FEATURES

Primary highway			
Secondary highway			
Light duty road			
Unimproved road	=====	=====	=====
Trail			
Dual highway			
Dual highway with median strip			
Road under construction			U. C.
Underpass; overpass			
Bridge			
Drawbridge			
Tunnel			

BUILDINGS AND RELATED FEATURES

Dwelling or place of employment: small; large	▪ ▨	▪ ▨	▪ ▨
School; church	i i	i i	i i
Barn, warehouse, etc.: small; large	□ ▨	□ ▨	▪ ▨
House omission tint			
Racetrack	⬭	⬭	⬭
Airport	✕	✕	✕
Landing strip	=====	=====	=====
Well (other than water); windmill	○ ⵐ	○ ⵐ	○ ⵐ
Water tank: small; large	● ⊘	● ⊘	● ⊘
Other tank: small; large	● ⊘	● ⊘	● ⊘
Covered reservoir	⊘ ▨	⊘ ▨	⊘ ▨
Gaging station	⚲	⚲	⚲
Landmark object	○	○	○
Campground; picnic area	⹁ ⊼	⹁ ⊼	⹁ ⊼
Cemetery: small; large	⊞ Cem	⊞ Cem	⊞ Cem

145

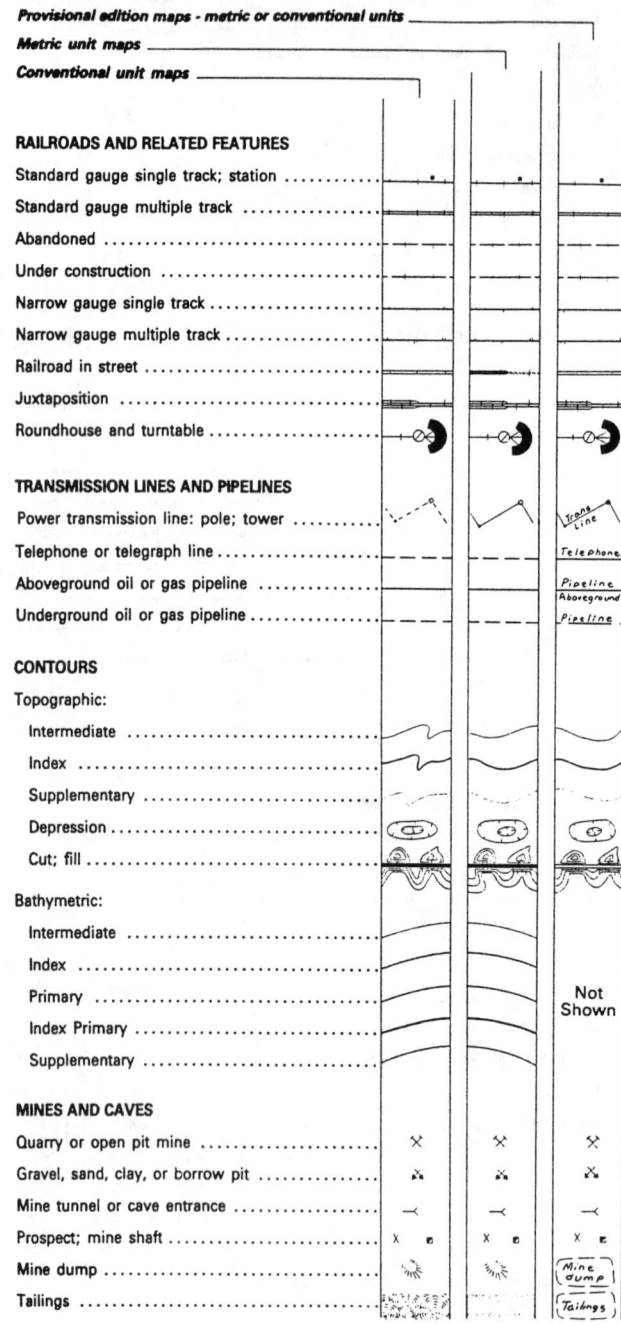

RAILROADS AND RELATED FEATURES

Standard gauge single track; station

Standard gauge multiple track

Abandoned

Under construction

Narrow gauge single track

Narrow gauge multiple track

Railroad in street

Juxtaposition

Roundhouse and turntable

TRANSMISSION LINES AND PIPELINES

Power transmission line: pole; tower

Telephone or telegraph line

Aboveground oil or gas pipeline

Underground oil or gas pipeline

CONTOURS

Topographic:

Intermediate

Index

Supplementary

Depression

Cut; fill

Bathymetric:

Intermediate

Index

Primary

Index Primary

Supplementary

MINES AND CAVES

Quarry or open pit mine

Gravel, sand, clay, or borrow pit

Mine tunnel or cave entrance

Prospect; mine shaft

Mine dump

Tailings

SURFACE FEATURES

Levee

Sand or mud area, dunes, or shifting sand

Intricate surface area

Gravel beach or glacial moraine

Tailings pond

VEGETATION

Woods

Scrub

Orchard

Vineyard

Mangrove

MARINE SHORELINE

Topographic maps:

Approximate mean high water

Indefinite or unsurveyed

Topographic-bathymetric maps:

Mean high water

Apparent (edge of vegetation)

COASTAL FEATURES

Foreshore flat

Rock or coral reef

Rock bare or awash

Group of rocks bare or awash

Exposed wreck

Depth curve; sounding

Breakwater, pier, jetty, or wharf

Seawall

BATHYMETRIC FEATURES

Area exposed at mean low tide; sounding datum

Channel

Offshore oil or gas: well; platform

Sunken rock

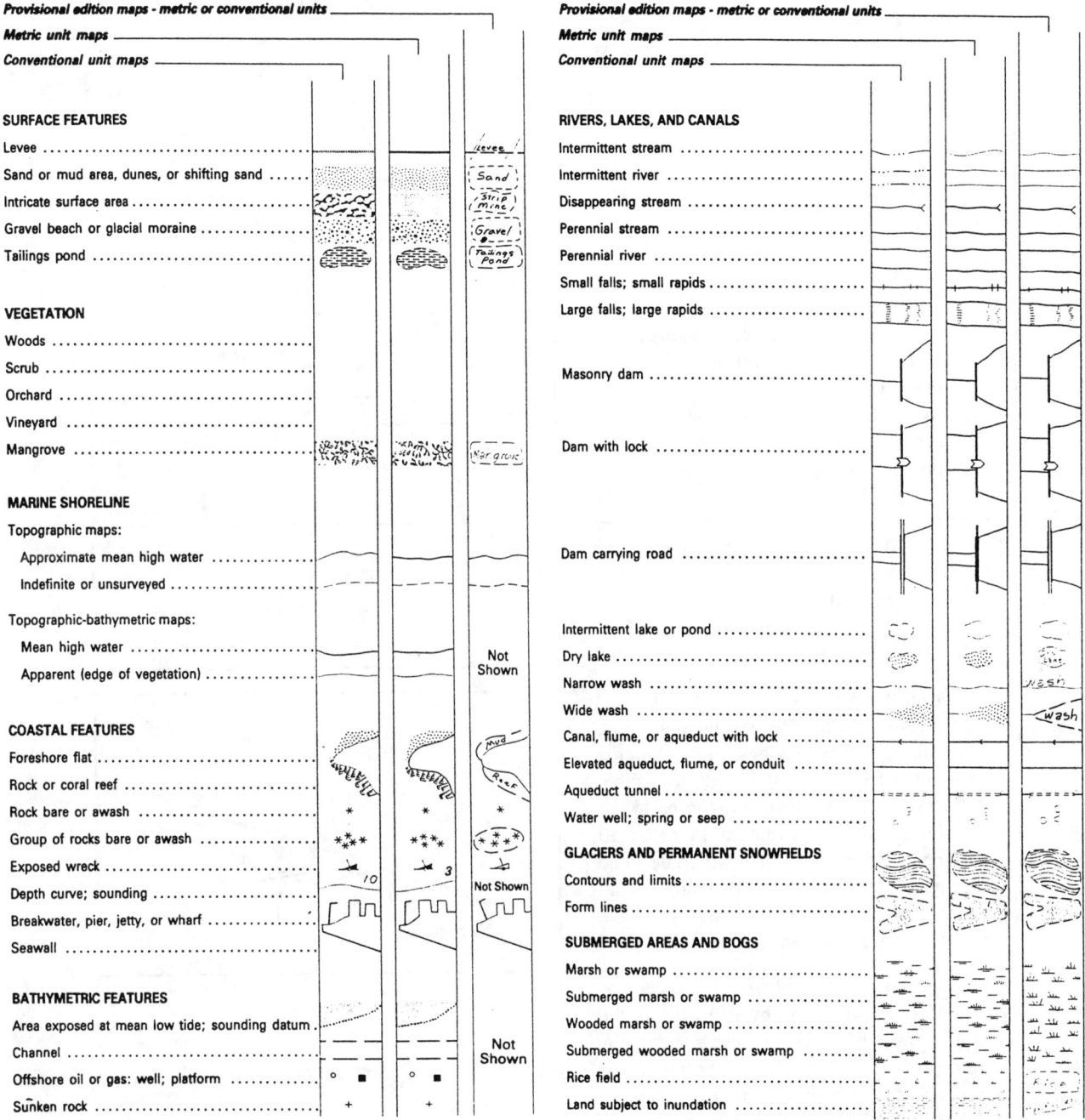

Provisional edition maps - metric or conventional units
Metric unit maps
Conventional unit maps

Provisional edition maps - metric or conventional units
Metric unit maps
Conventional unit maps

SURFACE FEATURES

Levee ....

Sand or mud area, dunes, or shifting sand ......

Intricate surface area ....

Gravel beach or glacial moraine ....

Tailings pond ....

VEGETATION

Woods ....

Scrub ....

Orchard ....

Vineyard ....

Mangrove ....

MARINE SHORELINE

Topographic maps:

  Approximate mean high water ....

  Indefinite or unsurveyed ....

Topographic-bathymetric maps:

  Mean high water ....

  Apparent (edge of vegetation) ....

COASTAL FEATURES

Foreshore flat ....

Rock or coral reef ....

Rock bare or awash ....

Group of rocks bare or awash ....

Exposed wreck ....

Depth curve; sounding ....

Breakwater, pier, jetty, or wharf ....

Seawall ....

BATHYMETRIC FEATURES

Area exposed at mean low tide; sounding datum .

Channel ....

Offshore oil or gas: well; platform ....

Sunken rock ....

RIVERS, LAKES, AND CANALS

Intermittent stream ....

Intermittent river ....

Disappearing stream ....

Perennial stream ....

Perennial river ....

Small falls; small rapids ....

Large falls; large rapids ....

Masonry dam ....

Dam with lock ....

Dam carrying road ....

Intermittent lake or pond ....

Dry lake ....

Narrow wash ....

Wide wash ....

Canal, flume, or aqueduct with lock ....

Elevated aqueduct, flume, or conduit ....

Aqueduct tunnel ....

Water well; spring or seep ....

GLACIERS AND PERMANENT SNOWFIELDS

Contours and limits ....

Form lines ....

SUBMERGED AREAS AND BOGS

Marsh or swamp ....

Submerged marsh or swamp ....

Wooded marsh or swamp ....

Submerged wooded marsh or swamp ....

Rice field ....

Land subject to inundation ....

CONSTRUCTION OF CROSS - SECTIONS

Illustration of one approach 1. For the selected station, indicate the location of the cross-section on the plan. A distance 30 m each side of the centerline should be adequate in most cases. This will fit lengthwise onto a letter sized page if a horizontal scale of 1 cm to 5 m is used. Using a vertical scale of 1 cm to 5 m continue as indicated below. *	Location of Cross-section 210 2 13 216 Plan view
2. Draw the centerline of the cross-section vertically on the center of the page, and locate on it: (a) the elevation of the ground at the centerline, determined by the examination of the contour lines and visual interpolation, or by measurement where preferred. (b) the elevation of the highway crown, by scaling from the profile completed earlier, or by computation.	Highway Centerline El., m 216 213 — Ground Elevation 210 Centerline Elevation (crown) Cross-section
3. Construct the line of the ground level along the cross-section by plotting the horizontal distance for each contour elevation from the highway's centerline and connecting the points.	El., m 216 213 210 Ground Line
4. From the point of the crown, construct the lines indicating the pavement surface, shoulders, slopes, ditches, embankments and other features of the cross-section as illustrated earlier. Also, show to scale the thickness of the pavement in order to define the surface of the subgrade. The surface, in turn defines the limits of the subgrade - excavation in a cutting, or the limits of the fill for an embankment.	Pavement Thickness Subgrade
Note the use of templates may speed the work, because the highway as far as the ditch slopes has a uniform cross-section. A template may be constructed from cardboard or thin plastic and may be varied to reflect the various cross-sections in cuttings and embankments.	Basic template, and ditch and slope varaitions
* Other scales may be used, depending on required accuracy and selected page size.	Sketches shown here are not to scale.

McDraw II, Highway graphic

Index

Stopping sight distance, 27, 31. *See also* Alignment; Speed

Surveys
route, 2
reconnaissance, 2
soil, 3

Topography
aerial photography, 1 - 3
effects on design, 2, 12
examination of, 2 - 8, 53, 81
physical features, natural and man-made, 1 - 12, 95
terrain, 1 - 12, 23, 31
topographic maps, 1 - 3

Traffic
average daily traffic (ADT), 20
design designation, 25
design hourly volume (DHV), 20
directional distribution (D),
K factor, 20
level of service guidelines, 20
volumes
related to design speed, 25
related to roadside design, 45

Traffic control devices
markings, 42
signs, 42

Trucks
climbing lanes, 31
curb radii requirements, 52
effects of grades on speeds, 31
percentage in traffic flow, 20

Unit prices for construction. *See* Construction

United States Geological Survey (USGS), 1, 2,

United States Soil Conservation Service, 2

Vertical curves. *See* Alignment